Protection against Erosive Wear Using Thermal Sprayed Cermet

Carlos P. Bergmann and Juliane Vicenzi

Protection against Erosive Wear Using Thermal Sprayed Cermet

A Review

 Springer

Authors

Dr. Carlos P. Bergmann
Universidade Federal do
Rio Grande do Sul
Escola de Engenharia
Depto. Materiais
Av. Osvaldo Aranha 99-7 Andar
90035-190 Porto Alegre Rio
Grande do Sul
Brazil
Telephone: 5133163405

Dr. Juliane Vicenzi
Universidade Federal do
Rio Grande do Sul
Escola de Engenharia
Depto. Materiais
Av. Osvaldo Aranha 99-7 Andar
90035-190 Porto Alegre Rio
Grande do Sul
Brazil
Telephone: 5133083637

ISBN 978-3-642-43045-9   ISBN 978-3-642-21987-0 (eBook)

DOI 10.1007/978-3-642-21987-0

Typeset & Cover Design: Scientific Publishing Services Pvt. Ltd., Chennai, India.

Printed on acid-free paper

9 8 7 6 5 4 3 2 1

springer.com

Preface

The erosion caused by solid particles under extreme conditions (high hardness and velocity of the erodent, cyclic load, and high temperatures) is problematic for industrial equipments. As function of requests, the tension generated in the particles and/or in the target material is, approximately, one order of magnitude greater than its mechanical strength, and as a result, the material can be damaged.

Many industries have invested in technology to minimize damage caused by degradation of the materials under erosive wear. In industrial environment, this type of wear occurs when abrasive particles are mixture or carried through pipelines, fans and cyclones, or, in equipments such as mills and mixers, by the inherent displacement of the load. This type of wear was found in mineral processing and separation industries. At high temperatures, for example, these processes take place in the energy-generation industry (thermal plants), in steel and cement production, paper production, and in the petrochemical industry.

The erosion at high temperatures leads to the deterioration of parts or components of machines, turbines, engines, and boilers that operate with particulate flows and, as a consequence, shortens their useful lifespan. In the petrol industry, for example, the erosion in boilers powered by powder fuel contributes to approximately 25% of the failures of these equipments, attributed to erosive wear caused by fly ashes.

Regarding erosion at high temperatures, several authors mentioned earlier suggest different regimens that contribute to damage of the material under erosion–oxidation, but there is no accordance among the findings by these authors. It is known that different variables in the process of erosive wear can lead to damage of the material by different mechanisms. However, due to the effect of temperature on the material, an oxide is formed, establishing the wear mechanisms. This mechanism result from erosion of the oxide and/or erosion of the composite (formed by oxide and substrate), making the approach for determining the wear phenomenon more complex.

The erosion of cermets (bulk and coating), more specifically at high temperatures, has been the subject of many researchers, although few of these studies are conclusive, and even fewer, agree with each other. In the case of cermet coatings, the complex microstructure, due to the lamellar formation can make even more difficult the understanding of the phenomenon responsible for erosion. Besides, as Stack and Pena (1997) noticed, in their studies of an alloy Ni13%Cr with WC particles at temperatures as high as 650°C, the change in the mechanical properties of the material due to temperature can contribute to increase in the complexity of the behavior of the wear of this kind of material.

The use of coatings such as cermets have shown excellent results with respect to the high strength against the erosive wear; however, very little is known about the mechanisms that lead to the degradation of this kind of material, at different work temperatures. This is attributed to the fact that they are not simple microstructures, formed by unique phases, or with properties of a bulk material, but are lamellar (formed due to the use of thermal aspersion technique) and complex microstructures, since each lamella is formed by a matrix phase (ductile), with carbides dispersed. The final properties are defined by the different phases present in the material, as well as the interaction among the lamellas.

Recent studies have suggested different regimens that lead to the damage of the material under erosion–oxidation, but there is no accordance among these studies. Besides, it is known that different variables in the erosive wear process (velocity, impact angle, flow rate of the particles, temperature, etc.) lead to damage of the material by different mechanisms (brittle, ductile, oxide erosion, erosion of the composite formed by oxide and substrate, etc.). In relation to composite material with metallic matrix, or coatings of this kind of material on metallic substrates, little research has been reported, or there are few conclusive studies with agreement among different authors (LEVY and WNAG, 1989; KUNIOSHI *et al.*, 2004; HULU *et al.*, 2005; FINNIE, 1995; HAWTHORNE *et al.*, 1999). This is probably due to the complexity of the wear behavior as function of the temperature and due to the modification in the mechanical properties of the material.

In this context, the authors present a review about possible mechanisms that are actuated and those that lead to degradation of bulk and, more specifically, coating cermets at different temperatures. This book is divided in 7 chapters covering subjects about coatings cermets production and erosive wear in different types of materials.

Contents

Protection against Erosive Wear Using Thermal Sprayed Cermet: A Review

Abstract. Erosive wear is characterized by successive loss of material from the surface due to the continuous impact of solid particles. This type of wear affects numerous industries, such as power generation, mining, and the pneumatic transportation of solids. The worst case scenario normally occurs where there is a combination of both erosion and oxidation, especially at high temperatures. In order to minimize damage caused by erosive wear, many authors propose the use of better bulk materials or surface coatings, and generally cermets (ceramic carbides in metal matrices) are suggested. Various researchers have conducted experiments to study the wear mechanisms occurring in this kind of materials, but most of these experiments do not lead to similar results; in fact, there is no accordance among the authors, and moreover, some wear variables are ignored. In this book, studies undertaken in this field by several investigators have been discussed extensively. At the end of this book, table reviews are suggested to summarize the most important mechanisms of the erosive wear in bulk and coating cermets.

Keywords: cermets, coatings, erosive wear.

1 Introduction

Erosion, according to the definition of authors such as Kulu (1989), Kulu *et al.* (2000), and Kulu and Halling (1998) *apud* Kulu *et al.* (2005), occurs when solid particles with high hardness and velocity strike against the surface of some material. As a result, the material tends to be damaged, depending on its structural characteristics and associated properties. In case of brittle materials, the brittle fracture is dominant, whereas for ductile materials the mechanism of micro-cutting and/or low cycle fatigue prevails. At high temperature, there occurs not only physical degradation of the material but also chemical degradation caused by oxidation (erosion) (ROY *et al.*, 1998 *apud* KUNIOSHI, *et al.*, 2004).

Brittle materials are more resistant to erosion at low attack angles, whereas ductile materials have better resistance at high attack angles (90°). When more than one of these situations are actuated, that is, when the material is removed by a combination of cutting, ploughing, brittle fracture, and fatigue mechanisms, there is a compromise between hardness and fracture tenacity of the materials.

The combination of different materials in a microstructural scale results in materials with peculiar properties and improved performance, which are not shown by the individual constituents. On the basis of this conception of composite materials, the cermets, which by the set of properties, show excellent erosion performance.

The concept of a metal matrix composite reinforced by particulates with high hardness is a natural consequence of the study of erosion of materials. The cermets (ceramic carbides in metal matrices) are widely used and are designated: WC-Co, Cr_3C_2/TiC-Ni, CrMo, WC-Ni, and Cr_3C_2-NiCr. These materials can be used not only as a bulk material but also as coatings over a matrix with poorer properties. In this case, they can extend the useful life of a component or even allow its operation under more severe condition, protecting the surface against corrosion (LEVY and WANG, 1988). In this sense, in addition to the alloys mentioned earlier, the self-fluxing Ni-based alloys (NiCrSiB) may also be used.

In order to apply these coatings over components, the different techniques in Surface Engineering is valuable. Qureshi and Tabakoff (1988), Tabakoff (1989), Burnet and Rickerby (1988), Jonsson *et al.* (1986) and Walsh and Tabakoff (1990) *apud* Shanov and Tabakoff (1996) employed processes such as plasma spray, sputtering, detonation gun, and electro-spark detonation to obtain protective coatings against erosion. Scrivani *et al.* (2001) employed the HVOF (*High Velocity Oxigen Fuel*) technique to obtain very hard coatings with excellent adhesion and cohesion properties. Coatings applied by arc-spray technique have also been used; however, this technique resulted in a coating characterized by large-sized lamellas and high porosity, limiting their performance (UNGER *et al.*, 1992, SOLOMON, 1997, VERSTAK *et al.*, 1998, WANG and LUER, 1994 *apud* WANG and SEITZ, 2001).

2 Coating Cermets Production

The utility of cermets against erosion wear has been known for 70 years. Different kinds of deposition process have been discussed in the literature (MATTHEWS and HOLMBERG, 2009), as is shown in Figure 1. Two important criteria for choosing the fabrication process are the thickness of the coating and the deposition temperature. According to Matthews and Holmberg (2009), as is shown in Figure 2, the typical range of the thickness varies between 0.1µm up to 10 mm, and the deposition temperature varies between 25°C up to 1000°C.

Attempts are being made to reduce the damage caused by erosion either by controlling the parameters that cause the erosion wear, or by using Surface Engineering techniques. A number of coating techniques like atmospheric plasma spray, High Velocity Oxyfuel (HVOF), thermal (TS) and detonation spray (DS) have been used over the years to deposit such coatings. Other coatings, such as oxides applied by atmospheric plasma spraying technique, cladding by laser process, spraying by detonation, PVD and boronising, plasma nitriding have been investigated in abrasive and erosive wear. It has been reported that detonation as well as HVOF sprayed coatings and boronising provide remarkable improvements compared to plasma nitriding in different velocities (MANN and ARYA, 2001).

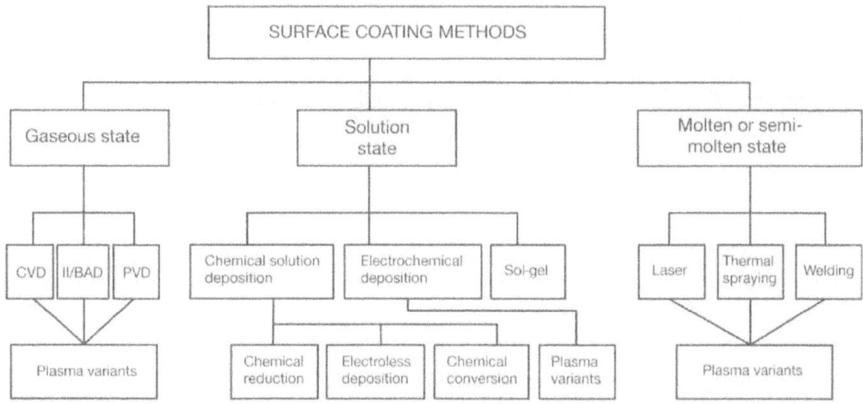

Fig. 1 A general classification of surface engineering techniques. (MATTHEWS and HOLMBERG, 2009). (CVD – chemical vapour deposition, II – ion implantation, IBAD – ion beam assisted deposition, PVD – physical vapour deposition).

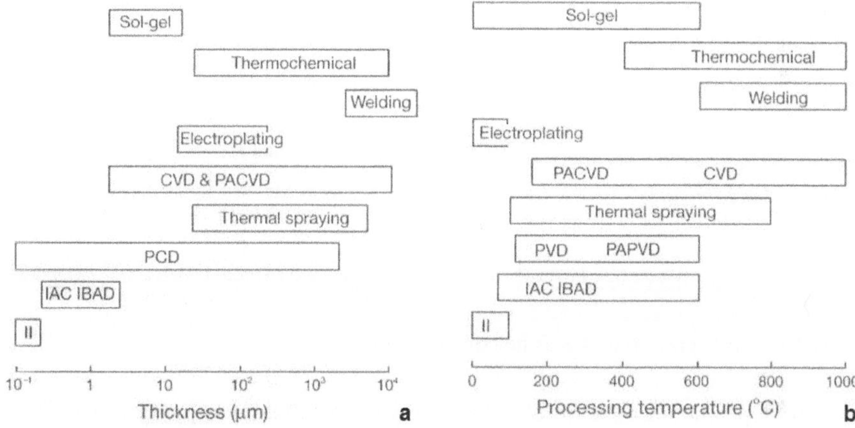

Fig. 2 Typical ranges for (a) depths of surface modifications and thicknesses of coatings, and (b) processing temperatures for coatings technologies. (MATTHEWS and HOLMBERG, 2009). (CVD – chemical vapour deposition, II – ion implantation, IBAD – ion beam assited deposition, PVD – physical vapour deposition, PA – plasma assisted, IAC – ion assisted coating).

Acoording to Shanov, Tabakoff and Gunaraj (1997) previous works demonstrated the excellent protection that CVD coatings provided for cemented tungsten carbide, for ceramic substrates, and for super alloys in particulate flow environments. In another erosion test they compared the CVD titanium carbide coating and ion nitriding treatment applied in two different substrates: INCO 718 and on AISI 410 (stainless steel 410). It was found that the ion nitriding treatment in glow discharge plasma does not improve the wear performance of the studied substrate materials. The CVD titanium carbide coating on both metals behaves as a brittle material and its erosion resistance increases significantly at elevated temperatures. This study demonstrated that the CVD titanium carbide coating provides excellent erosion protection for INCO 718 and stainless steel 410 when subjected to impact by chromite particles at elevated temperatures.

2.1 Thermal Spray Processes

In Surface Engineering, thermal spray is a generic term for a group of manufacturing processes whose purpose is to get into any kind of substrate metal, ceramic, polymer or composites coatings. Thermal spraying is a process in which molten, semi-molten or solid particles are deposited on a substrate. Consequently, the spraying technique is a way to generate a stream of such particles. The coatings are produced by the energy source. The energy sources are used to heat a feed coating material to a molten or semi-molten state. Coatings can be generated if the particles can plastically deform at impact with the substrate, which may only happen if they are molten or solid and sufficiently rapid. These processes are classified according to the energy source, chemical or electrical, as presented in Figure 3.

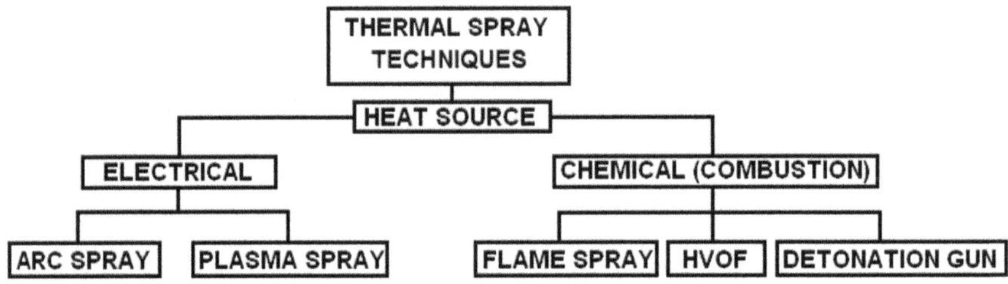

Fig. 3 Thermal spray processes as function of energy source.

The deposition occurs through the injection of a material as powder, wire or rod into a heat source (gas combustion, electric arc or plasma - Table 1). The main element that these processes have in common is that they all use a heat source to convert powders or wires into a spray of deformable particles (molten or sometimes semi-molten). The particles are accelerated by the expansion of gases and are projected with high speed on a substrate, causing a high energy impact, and cool down rapidly, producing a coating with several layers of fine, overlapped particles (Figure 4). Upon impact, a bond forms with the surface and subsequent particles cause thickness buildup. The coatings obtained by this technique have a thickness ranging from micrometers to millimeters.

Table 1 Flame temperature for different thermal spray processes. (SULZER METCO, 2005).

Flame temperature (K)			
Combustion (wire or powder)	HVOF	Electric arc (wire)	Plasma
3000	2000-3000	4000-6000	12000-16000

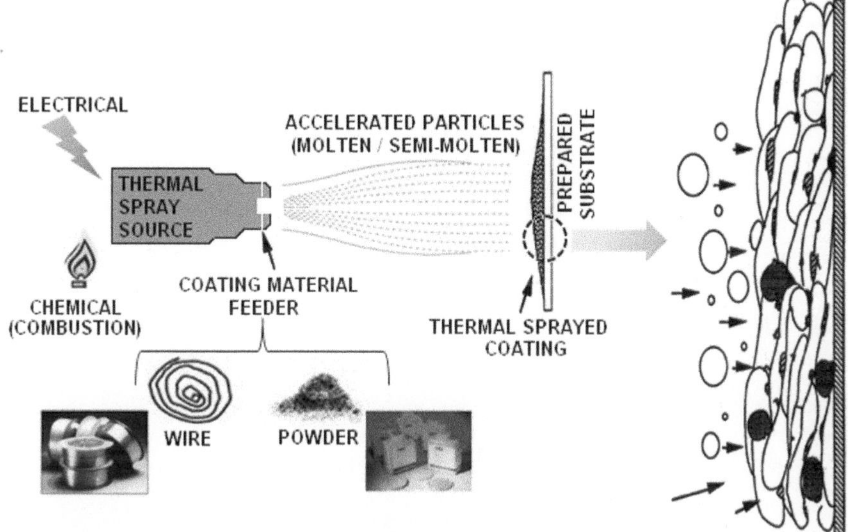

Fig. 4 Schematic diagram of lamellae formation of coatings on substrate.

According to ASM (2004), there are advantages and disadvantages to using the thermal spray process. Among the main advantages cited: any material capable of melting without decomposing can be used. Moreover, coatings over any substrate and a good control of temperature during the deposition process can be applied. On the other hand, disadvantages are about the limitations in size, the impossibility to cover small and / or deep cavities in which the torch or gun cannot fit. Besides the classification presented in Figure 3, the thermal spray process can be divided according to kinetic energy, controlled atmosphere, productivity and the possibility to deposit nanostructured materials. In addition, each process has different cost, materials flexibility and coating performance capabilities.

Currently, the thermal spray coatings are of high enough quality and some industries use these technologies, considering that these coatings have been used for over 100 years. For example, the key aircraft engine components and biomedical prostheses are routinely coated using thermal spray technologies. Many industrial components can be protected and have increased, extended or enhanced their shelf life using thermal spray. Hard, wear-resistant, coatings are used in automotive engines, insulators are sprayed, chemical reactors are repaired against corrosion, pumps are restored, bridges are coated and aircraft bodies and engine parts are protected.

The spraying techniques can be divided into different forms; one of them has been shown in Figure 3 (as a function of energy source). Another one uses medium velocity or medium temperature or type of feeding material in the thermal spray process. Figure 5 shows the processes as a function of temperature and material delivery speed.

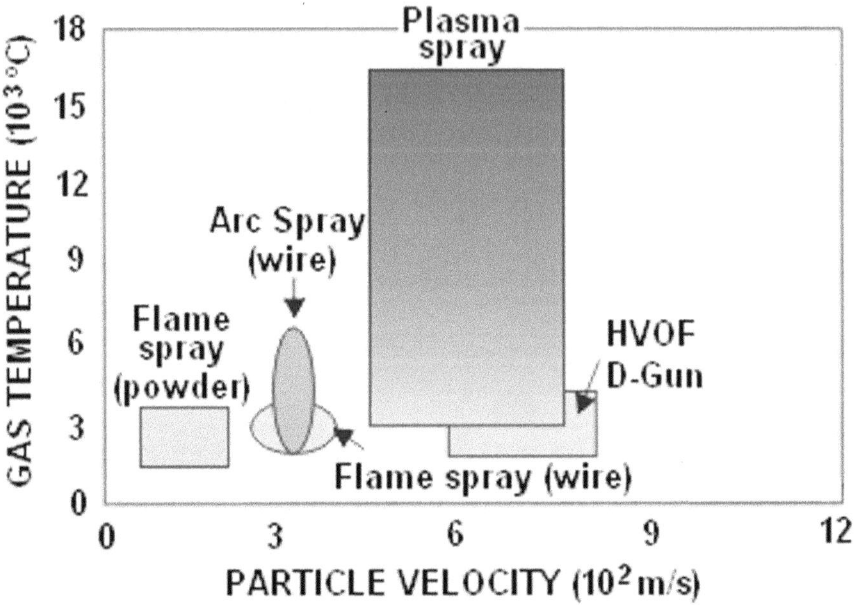

Fig. 5 Thermal spray processes as function of the processing temperatures and material transport velocities. (Adapted by GROSS, 2001).

The focus of this book is on the techniques that are used to deposite cermets. Cermets are intensively used in thermal spraying industry for applications where wear resistance is required. According to Souza and Neville (2003) *apud* Powlowski (1994) high velocity oxy-fuel (HVOF) process has been used to produce cermets coatings with low porosity (typically <1%). In addition, the authors related that the use of higher exhaust velocities and lower flame temperatures than with other processes, avoids alteration of the mechanical properties of the substrate, coating matrix or hard phase particles.

The WC-Co is the most widely applied composite according to Berger *et al.* (1996), nevertheless, there are limits to the possible applications due to the tungsten oxidation in elevated temperatures. At higher temperatures corrosion and oxidation can occur simultaneous, in these cases the superior oxidation response of Cr_3C_2 enables Cr_3C_2-NiCr coatings to be applied. Some features of this cermet (high melting point, maintains high hardness at elevated temperatures, strength and wear resistance up to maximum operating temperatures approaching 900°C) have lead this coating to be applied industrially to mitigate the effects of high temperature erosion in turbine based power generation applications (MATTHEWS, JAMES and HYLAND, 2009).

Under erosion conditions at high temperature Matthews, James and Hyland (2009) *apud* Wang and Luer (1994) and Sue and Tucker (1987) observed that chromium carbide based coatings applied by HVOF or D-Gun show dissimilar behaviors. From 25°C to 300°C an initial reduction in erosion rate was observed when temperature increased, beyond this the erosion rate increased at a constant rate with increasing temperature up to 750°C, by HVOF coating. In D-Gun coatings was observed a gradual increase in erosion rate with temperature up to 500°C.

The kind of feedstock is also responsible for the erosion resistance due to the characteristics of the powder based coatings and the effect of deposition technique. For example, Matthews, James and Hyland (2009) reported that by agglomerated/sintered powder based coatings the effect of deposition technique was most evident at 800°C, due to the variation in carbide dissolution in-flight. The HVOF coating suffered the greatest degree of carbide dissolution leading to a more ductile erosion compared to the HVAF coating, richer in carbide.

According to Stoica *et al.* (2004), the latest HVOF deposition systems are designed to optimize the velocity and temperature of the spray particles, hence decreasing the level of in-flight chemical reactions, and improving the bonding throughout the coating. A survey of the published literature of such systems shows that the resulting coatings are more durable in wear applications when compared to the older HVOF or atmospheric plasma systems (APS).

Besides the technique used to deposit the coatings, the coating design is also responsible for their tribomechanical properties. For example, the composition and the structure of the coating, plays a dominant role in controlling the tribomechanical properties of the coating. Thus, functionally graded depositions which bring a gradual change in both the composition and structure have been shown to alter the coating residual stress and thus to decrease the fatigue and fracture failures, leading to higher wear resistance (STOICA *et al.*, 2004).

There are many thermal spray processes and variations of thermal spraying technology currently available for depositing coatings. For example, in flame spraying the combustion of a fuel gas is used to heat the material. In atmospheric plasma spraying (APS) the material is melted and accelerated in a plasma jet. To avoid oxidation of the feed material, spraying can be carried out in an inert gas atmosphere, at a reduced pressure (known as vacuum plasma spraying VPS or low pressure plasma spraying LPPS). In high velocity oxy-fuel spraying (HVOF), material is injected into a high velocity jet generated by burning a fuel mixed with oxygen at high pressure.

The thermal spray processes present differences in the final coating. The supersonic velocity of the HVOF spray gives a thin, dense and well bonded splat and variants of the gun are being developed for many applications. Other traditional methods melt two wires in an arc or feed powder into a high speed plasma gas stream and, for high performance applications, use a low pressure or inert gases. Some new developments include radio-frequency induction plasma with its high temperatures or the opposite process with a high velocity, low temperature, localized spray which gives pure, dense material with high compressive stresses.

Of these thermal spray process cited, the high velocity oxy-fuel (HVOF) thermal spray process has come to be widely used to produce wear-resistant components. A major advantage of this approach is that it can yield cermet-based coatings having a lower level of porosity than other traditionally used thermal spray processes such as arc spraying or conventional plasma spraying. There is also a relatively new plasma spraying process, called high power plasma spraying (HPPS) that enables the production of cermet-based coatings having a relatively low level of porosity. One of the advantages of the HPPS process is that comparatively high deposition efficiencies are possible when depositing many types of cermets. (VOYER and MARPLE, 1999). On the other hand, in some applications the porosity of coating can be necessary. For example, in a corrosion environment the coating can produce a oxide film protector. In this case the plasma spray process is a good choice. Another example, according to Voyer and Marple (1999) can be found in sliding wear applications. Porosity present in the thermal spray coatings appeared to play a key role in determining the wear resistance of the coatings during sliding contact with carbon-graphite. The more porous coatings produced using the HPPS process experienced lower wear than the denser coatings deposited using the HVOF process. It is believed that the pores in the coating serve as reservoirs for the debris generated during wear and trap the graphite-containing material, which serves as a lubricant.The aim of this book is to highlight the thermal spray process used to produce cermets coatings applied in erosion wear environments at different temperatures. As the results will show throughout this book, both thermal spray processes (plasma spray and HVOF) produce excellent coatings for use against erosion wear. Below, we will present these two, cited processes.

2.1.1 Plasma Spray

In plasma spraying process molten or heat softened material is accelerated onto a substrate to provide a coating. The plasma is the mechanism responsible for accelerating the particles. The plasma – a cloud of ionized gas with sub-atomic particles – is the result of a gas passing through an electric field of high intensity. Large amounts of energy are released, mainly by ultraviolet radiation. In this process, particles can reach speeds up to 300 m/s, within its trajectory in the flame, and extremely high temperatures of, 13,000 to 30,000K, depending on the power equipment.

The plasma spray process can produce coatings with range in thickness from a few micrometers to several millimeters. Plasma spray coating raw materials include metals, ceramics and a mixture of the two (both ceramic and metal), the cermets. Additionally, plasma spray coatings can be done in a wide array of conditions and it is an adaptable process.

According to Zatorski and Herman (1991), of the deposition methods, plasma spray is the only one that operates at high temperatures and the specific energy density is high enough to melt the materials, presenting a stable melted phase. Joshi (1992) adds that the temperature of the particles that are accelerated and heated, is influenced by particle injection speed.

Figure 6 shows schematically the process of plasma spraying. The powder is injected into a very high temperature plasma flame, where it is rapidly heated and accelerated to a high velocity. The accelerated particles impact on the substrate surface and rapidly cool forming a coating. The substrate temperature can be kept low during processing. The objective is to minimize damage, metallurgical changes and distortion to the substrate material.

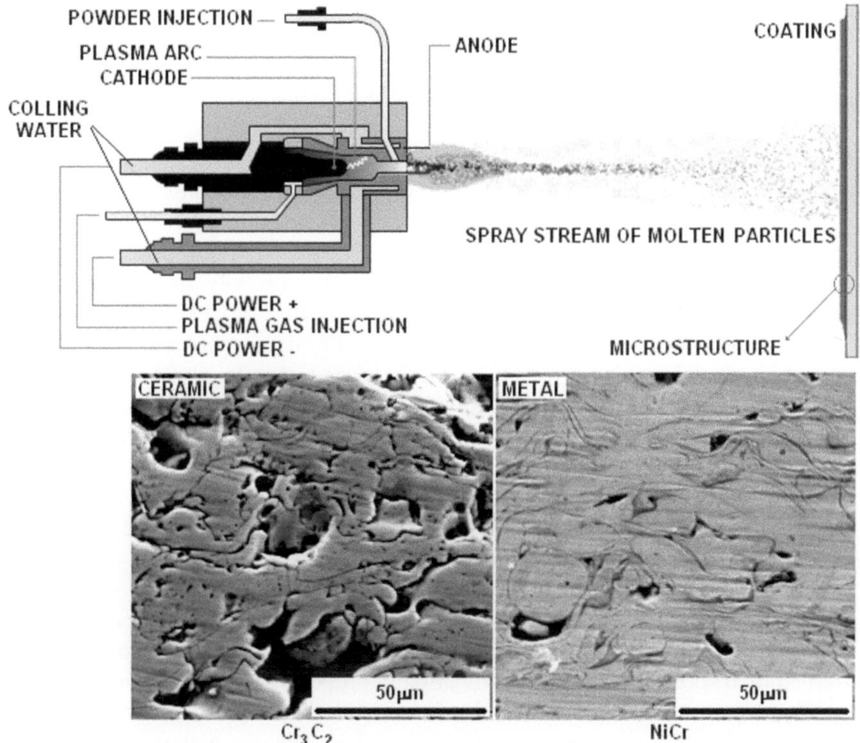

Fig. 6 Plasma spray processes: gun, plasma, molten particles and microstructure.

In Figure 6 the basic principles of a plasma spray torch operation can be observed. The plasma spray gun is made up by an anode and a cathode, both water cooled. The plasma begins with a high voltage discharge that causes ionisation in the gas and a conductive path for a DC arc to form between cathode and anode. The system reaches extreme temperatures forming a plasma. An inert gas – the plasma gas – (argon, nitrogen, hydrogen, helium) flows around the cathode and through the anode. The anode is shaped as a constricting nozzle. When the plasma is stabilized and is ready for spraying the electric arc extends down the nozzle. The powder is fed into the plasma flame most commonly via an external powder port mounted near the anode nozzle exit. The powder is so rapidly heated and accelerated that spray distances can be in the order of 25 to 150 mm. Nozzle designs and flexibility of powder injection schemes, along with the ability to generate very high process temperatures, enables plasma spraying to utilize a wide range of coatings. In addition, substrate temperatures may be controlled during deposition, making it possible to put a wide variety of plasma coatings on an equally wide variety of substrates.

The plasma spraying process is versatile in function of the raw material used as powder injection and the characteristics of the coatings obtained. The advantages of this process are: i) it can spray very high melting point materials, for example, refractory metals like tungsten and ceramics like zirconia; ii) the coatings obtained in general are much denser, stronger and cleaner than the other thermal spray processes (exception: HVOF and detonation processes); iii) there are many applications for the coatings due to the microstructure obtained: aerospace, automotive, medical devices, agriculture, communication, SOFC (Solid Oxide Fuel Cell), biomedical, erosion wear, abrasive wear, corrosion. Disadvantages of the plasma spray process are its relative high cost and the complexity of process.

The plasma spray process is most commonly used in normal atmospheric conditions and referred to as APS. Some plasma spraying is conducted in protective environments using vacuum chambers, normally back filled with a protective gas at low pressures; this is referred to as VPS or LPPS.

The plasma spraying process has been applied to deposit ceramic, metallic and cermets materials. According to Nerz, Kush and Rotolo (1991), the residual porosity is the critical characteristic of plasma spray coatings. This porosity cannot be totally annulled, but may be influenced by operational parameters of deposition.

Regarding the texture of the surface being coated, Ladsdown and Price (1986) emphasize that this depends on the initial roughness of the substrate, the plasma power and the particle size of powder.

Sevostianov *et al.* (2004) developed a quantitative study to characterize microstructures of plasma spray coatings with the properties of these materials. From this study, the authors concluded that the porosity and the presence of cracks are mainly responsible for the modification in properties of plasma spray coatings, mainly due to anisotropy in the distribution of these defects, which produces anisotropic elastic and conductive properties. As a rule, the authors suggest that a porosity of less than 15% affects the elastic and conductive properties of the material less than the micro cracks present in the coatings (due to the shape and random distribution).

On the other hand, levels above this can also be indicators of high crack densities, leading to the modification of properties. The control of porosity and micro cracks obtained after deposition of plasma spray coatings are not the only ones responsible for the modification of the microstructure modification. Due to the high temperatures to which the powders are subjected, as well as the gases used to generate the plasma, there may be changes in their crystalline structure, and during particle impact and solidifying new phases may be present.

The microstructure of the plasma spray coatings depends also on the equipment used and on the parameters employed during the spraying process. For example, Staia *et al.* (2001) evaluated the production of different Cr_3C_2-25% NiCr phases due to modifications in spray parameters. For example, the gas responsible for generating the plasma (using nitrogen) and the gas pressure during the spraying (300 to 1200mbar) were varied. The authors observed the powder feeding crystalline structure was NiCr face-centered cubic (FCC) and orthorhombic Cr_3C_2. After spraying with nitrogen, beyond the original powder phases, carbo-nitride phases were found, depending on the C/N ratio, such as $Cr_{6,2}C_{3,5}N_{0,3}$ and $Cr_3C_{1,52}N_{0,48}$. After spraying in the atmospheric pressure (APS), the phases Cr_7C_3 and CrO_2.were also identified Tables 2 and 3 summarize the conditions used to spray and the properties obtained for the coatings.

Table 2 Plasma spray thermal conditionsof the NiCr-25Cr_3C_2 (STAIA *et al.*, 2001).

Condition	Spraying pressure (mbar)	Plasma gas flow: $N_2/He/H_2$ (SPLM)	Spraying distance (mm)	Power imput (kW)	Carrier gas flow: N_2 (SPLM)	Gas Cooling (N_2)
1	300	50/15/4	120	39	1,5	No
2	300	50/15/4	120	39	1,5	Yes
3	1200	46/15/3	120	35	2,0	Yes
4	1200	46/15/3	140	35	2,0	No
5	1200	46/15/3	120	35	2,0	No
6	1000 (APS)	50/10/3	120	35	2,0	Yes

Staia *et al.* (2001) observed that the volume fraction phase Cr_3C_2 + carbo-nitride phase enhances with the increase of spray pressure, the decrease of spray distance and the absence of cooling gas. The increase of graphite carbon in the carbo-nitrede phase was observed in the coating deposited with the highest pressure. The highest hardness was obtained for the phase containing carbides, spraying with 1200mbar pressure, with a distance of 120mm and pre-heating of the substrate at 600° C.

Table 3 Properties obtained after plasma spray NiCr-25Cr$_3$C$_2$ in the conditions cited in Table 2 (STAIA *et al.*, 2001).

Condition	V$_f$ Ni-Cr (%)	V$_f$ Cr$_3$C$_2$ + carbo-nitride phase	Eficciency Cr$_3$C$_{2coating}$/ Cr$_3$C$_{2powder}$	V$_f$ porosity + 2^a phase	HV$_{50}$NiCr phase	HV$_{50}$ carbeto phase
1	66,5	31,3	41,7	2,2	-	-
2	77,8	18,9	25,2	3,3	301,6	1468
3	36,0	58,2	77,6	5,8	255,2	1866
4	41,2	53,2	70,9	5,6	264,5	2115
5	30,2	64,8	86,4	5,0	277,5	2296
6	38,4	54,8	73,0	6,8	217,3	1445

V$_f$ = volumetric fraction.

Zhang *et al.* (2009) evaluated the use of different guns for producing Cr$_3$C$_2$ – NiCr plasma spray coating to be applied in resistance cracks fatigue (RCF). For the conventional air PS guns, a disordered plasma-jet generated with a ''boundary effect'' was observed. The authors noted that as the distance from the nozzle exit increases, the enthalpy decreases, consequently, conventional air PS limits coating quality with low bonding strength and high porosity.

The other gun used was a development by the National Key Laboratory for Remanufacturing, China, a high-efficiency plasma spraying (PS) system with a hypersonic PS gun. Using the PS gun, the velocities of the plasma jet, and particles can achieve 2400 m/s and 500 m/s at the normal distance of 100 mm from the nozzle exit, according to Zhang *et al.* (2009) *apud* Zhu, Xu and Yao (2005). When compared with the conventional air PS, the quality of the coatings deposited by the high efficiency hypersonic PS improves greatly. This can be seen in Figure 8, the cross-sectional image of a coating microstructure with a back scattered electron model deposited by the high efficiency hypersonic PS. A few intra-lamella cracks (as indicated by the arrow) and pores at the boundaries of splats are visible in the coating. This microstructure (Figure 7) can be compared with microstructures obtained for Vicenzi (2007) using a conventional air PS gun (Metco 7MB) for Cr$_3$C$_2$-NiCr plasma sprayed coatings with 35% and 70% of carbides, as shown in Figure 8.

Figure 8 (a) and (b) shows the typical lamellar structures, with porosity and some particles of the plasma sprayed coatings not melting (or becoming plastic). The porosity evaluated by image analysis present values around 30% in both microstructures.

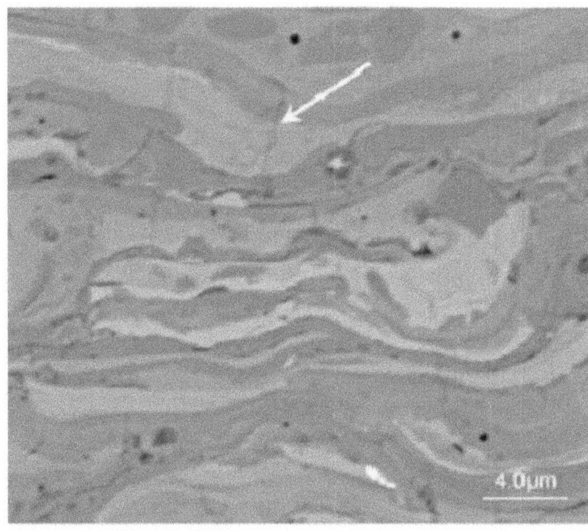

Fig. 7 Cross-sectional microstructure of plasma sprayed CrC-NiCr coating obtained by SEM microscopy (ZHANG *et al.*, 2009).

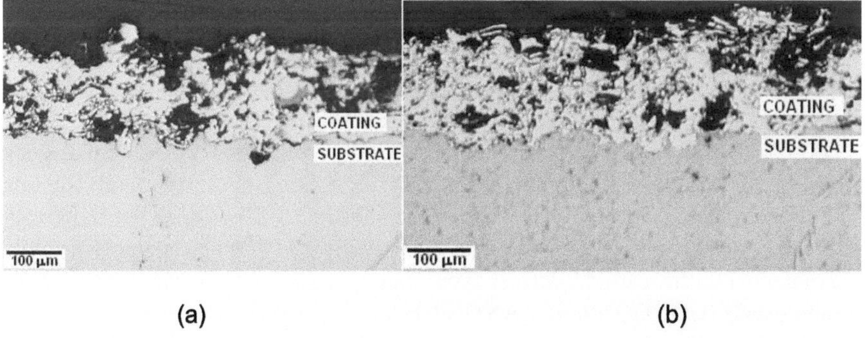

(a)   (b)

Fig. 8 Cross-sectional microstructure of plasma sprayed Cr_3C_2-NiCr coating obtained by optical microscopy (a) 35% and (b) 70% of Cr_3C_2 phase in the coating. (VICENZI, 2009).

2.1.2 HVOF (High Velocity Oxygen Fuel)

HVOF spraying has been considered advantageous when compared with other thermal spray processes, mainly for materials with melting points below 3000 K. The HVOF-coatings present higher density and bond strength, compared with others processes, making it attractive for many wear and corrosion resistance applications. According to Mann and Arya (2001) its high coating quality results from the use of a hot combustion-driven high-speed gas jet for thermal spraying. These coatings have environmental advantages compared to chemically/electrochemically formed coatings.

One of the differences between the HVOF and plasma spray process is the heat transferred to the substrate during spraying. Although the plasma temperature is higher than in the HVOF system, the heat transferred to the substrate by HVOF is usually greater than in the plasma spray process. The temperature gradient of the flame plasma is usually higher than the combustion flame of the HVOF, being less sensitive to changes in injection rate of powder particles when compared to plasma (JOSHI, 1992).

The HVOF utilizes confined combustion and an extended nozzle to heat and accelerate the powdered coating material. Typical HVOF devices operate at hypersonic gas velocities. The extreme velocities provide kinetic energy which help produce coatings that are very dense and very well adhered in the as-sprayed condition. There are a number of HVOF guns which use different methods to achieve high velocity spraying. The HVOF sprayed coatings are commonly applied by HP/HVOF JP-5000, DS-100, Met jet II, OSU, Diamond jet and Praxair 2000 HVAF systems. These systems are based on liquid as well as gaseous fuel and oxygen/air.

In the hypersonic spraying process (HVOF High Velocity Oxy-Fuel-), the burning of fuel with oxygen takes place inside a chamber in the spray gun. Fuel (kerosene, acetylene, propylene and hydrogen) and oxygen are fed into the chamber, combustion produces a hot high pressure flame which is forced down a nozzle increasing its velocity. Powder may be fed axially into the HVOF combustion chamber under high pressure or fed through the side of a laval type nozzle where the pressure is lower.

Due to the high pressure inside the combustion chamber, the expansion of gases can propel particles injected into the flame to speeds above 2000 m/s. Compared to plasma, however, the flame reaches much lower temperatures, for example, 2600°C using kerosene as fuel. Figure 10 sumarizes the HVOF thermal spray process, as well as the thermal sprayed HVOF coating microstructure. The very high kinetic energy of particles striking the substrate surface does not require the particles to be fully molten to form high quality HVOF coatings. This is an advantage for the carbide cermet type coatings and that is where this process really excels (GORDON ENGLAND, 2011).

The physical and chemical characteristics of the feeding powder have an effect on the heating and acceleration of the particles. With high-speed combustion, the discharge of the gun accelerates the feeding powder particles at a velocity higher than in plasma thermal spray equipment, around 2000 m/s. In both cases (HVOF and plasma) the particle size is the main parameter to determine particle plasticity and velocity during deposition.

Due to moderate temperatures employed in the hypersonic case, particles larger than a critical size do not melt completely. Therefore, uniformity of size distribution is a key factor for obtaining high quality coatings. The coatings obtained show high densification due to the intensity of the shocks of the particles with the substrate (see for example, the microstructure illustrated in Figure 9). As emphasized by Fillion (1995), HVOF-coatings have high adhesion, high densification and low residual stress in the coating.

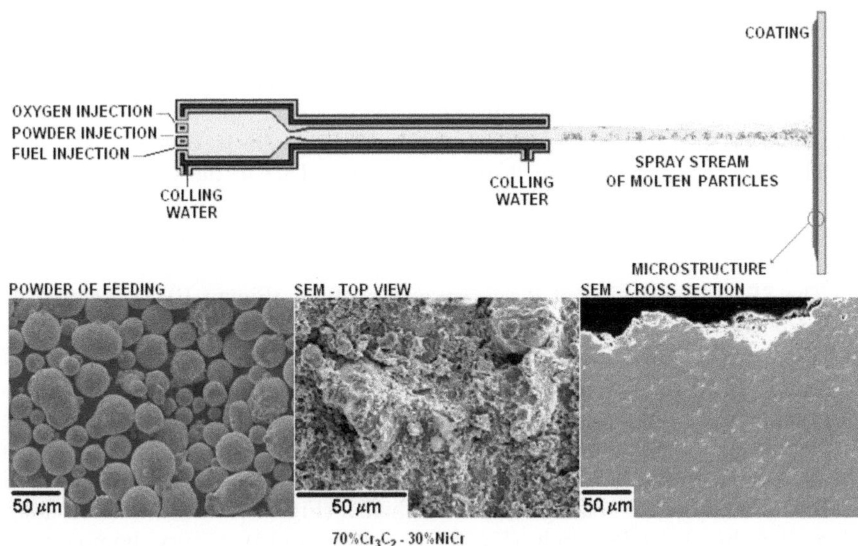

Fig. 9 Schematic diagram of HVOF thermal spray process (adapted from GORDON ENGLAND, 2011).

The properties of HVOF-coatings are highly dependent on a number of parameters including the preparation of the part surface, composition, morphology, size distribution and feed rate of the powder, and the precise control of gas flows, relative gun/part motion, stand-off angle of deposition and part temperature. Beyond that, the degree to which gas-powder reactions occur, depends on the specific device, operating parameters, and the material being deposited.

According to Souza e Nevile (2003) the HVOF process has been shown to produce good quality coatings of low porosity (typically <1%), avoiding alteration of the mechanical properties of the substrate, coating matrix or hard phase particles. Gordon England (2011) agrees with these authors. The coatings produced by HVOF are very dense, strong and show low residual tensile stress or in some cases compressive stress, which enable substantially thicker coatings to be applied.

However, in HVOF thermal spray coatings partial melting of the particles can occur causing poor bonding at the interfaces between the unmelted and semi-melted particles, beyond the phase composition changes due to in-flight chemical reactions. To mitigate these problems post-treatments in the coatings can be used, such as austempering/annealing, laser melting, and hot isostatic pressure (HIP). Promotion of cohesion between splats and adhesion between coating and substrate, phase transformation from amorphous to crystalline, generation of uniform compressive stress, and attenuation of coating anisotropy to obtain near homogeneous material properties are some of the desired post-treatment improvements, when a high degree of resistance to fatigue, impact loading, delamination, and corrosive/erosive/abrasive wear is required (STOICA, *et al*.2004).

There are a number of HVOF guns which use different methods to achieve high velocity spraying, Toma, Brandl and Marginean (2001) evaluated cermet coatings obtained with two different equipments: Top Gun and the OSU Carbide Jet System (CJS torch). The main difference between the spraying devices is the pressure in the combustion chamber. The CJS torch is operated at higher combustion chamber pressure than the Top Gun. Due to the CJS torch design (divergent conver-gent nozzle system), this system ensures a higher thermal efficiency with high particle velocities, producing dense coatings and reduces oxidation.

The SEM examination of the as-sprayed coatings (Figure 10) shows that the WC-Co and WC-Co-Cr coatings (deposited by CJS torch) have a more dense structure and a better distribution of the carbide particles than the WC-Cr$_3$C$_2$-Ni and Cr$_3$C$_2$-NiCr (deposited by Top Gun) sprayed coatings. Figure 10 (a) shows a very inhomogeneous structure with pure matrix areas in the WC-Cr$_3$C$_2$-Ni coating. The SEM micrograph of the Cr$_3$C$_2$-NiCr coating demonstrates that the distribution of the carbide particles in the matrix is more uniform than in the WC-Cr$_3$C$_2$-Ni, but the porosity of the coating seems to be higher (Figure 10 (b)). For the coatings deposited by the CJS torch (Figure 10 (d) and (e)) a very homogeneous microstructure was observed and the porosity was estimated to be lower than coatings deposited by Top Gun.

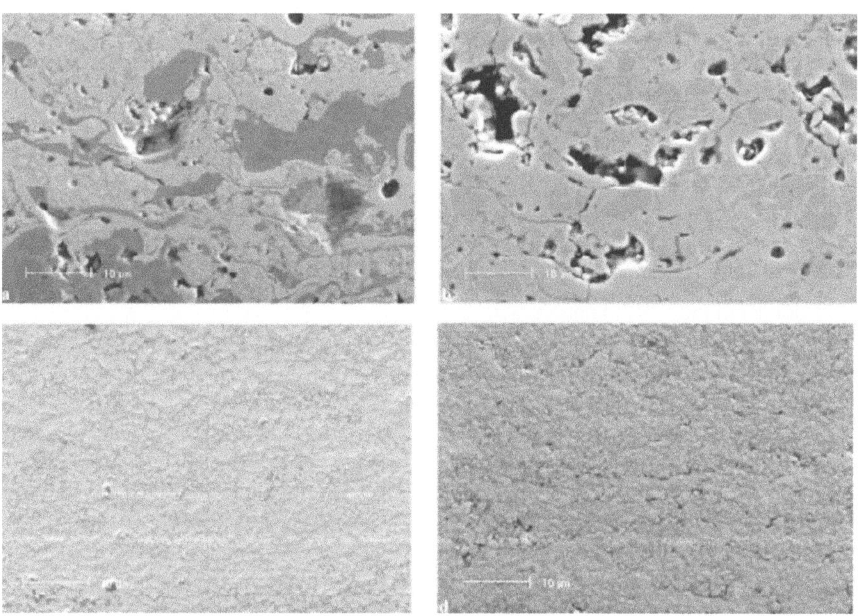

Fig. 10 SEM micrographs of the as-sprayed cermet coatings: (a) WC-Cr$_3$C$_2$-Ni, (b) Cr$_3$C$_2$-NiCr, (c) WC-Co and (d) WC-Co-Cr (TOMA, BRANDL and MARGINEAN (2001)).

The main application of HVOF coatings have been metal matrix composites (Ni, Cr, Co, or alloys of these elements) containing particles of carbides (WC, Cr_3C_2), which provide better wear resistance. Compared with the plasma spray process, the HVOF process maintains the integrity of the carbide particles. During the application and cooling of these coatings, chemical transformations can occur as a complex thermal decomposition of WC or Cr_3C_2 and reactions of carbides with the metal matrix (KARIM *et al.*, 1993 and MOHANTY *et al.*, 1996). The HVOF process promotes lower percentage of phase transformation and denser coatings with lower porosity, due to lower flame temperature and higher particle velocity.

HVOF coatings can be found in several and diverse industries, for example, industries needing wear resistance such as agricultural and construction equipment, food processing, aerospace, medical instruments that require high performance.

According to Mann and Arya (2001) HVOF is used to combat the erosion and corrosion occurring in hydro power plants and pumps. In applications where abrasive or erosive wear resistance is of primary importance, WC–Co with and without nickel or chrome is used. WC–Co–Cr powders are preferred when high corrosion resistance is needed.

2.1.3 Thermal Spray Coatings Microstructure

A typical microstructure of coatings obtained by thermal spraying is shown, schematically, in Figure 11. The sprayed coatings are produced by a process in which molten or softened particles are applied by impact on to a substrate. The particles become flattened and adhere to the surface irregularities after strike against the substrate (observe lamellar microstructure in Figure 11). A common feature of all thermal spray coatings is their lamellar grain microstructure resulting from the rapid solidification of small particles, flattened from striking a cold surface at high energy. In addition, in the microstructure pores, not blown particles and inclusions of oxides (in the case of coatings with metals) can be present. The bonding between the sprayed deposit and the substrate can be mechanical, metallurgical, chemical, physical or a combination of these forms (LIMA E TREVISAN, 2002).

The lamellar microstructure of HVOF-coatings produces anisotropic properties. For example, strength in the longitudinal direction can be 5 to 10 times that of the transverse direction (GORDON ENGLAND, 2011).

The morphology of the coating material depends of the raw material (powder feeding) such as: size, morphology and chemical uniformity of the powder particles. These powders can be obtained through the process of sintering, agglomeration, blending and fusion (HAWTHORNE *et al.*, 1997). These factors of the powder are correlated with the stream powder into the gun, injection in the flame and spread of highly plastic particles on the substrate surface.

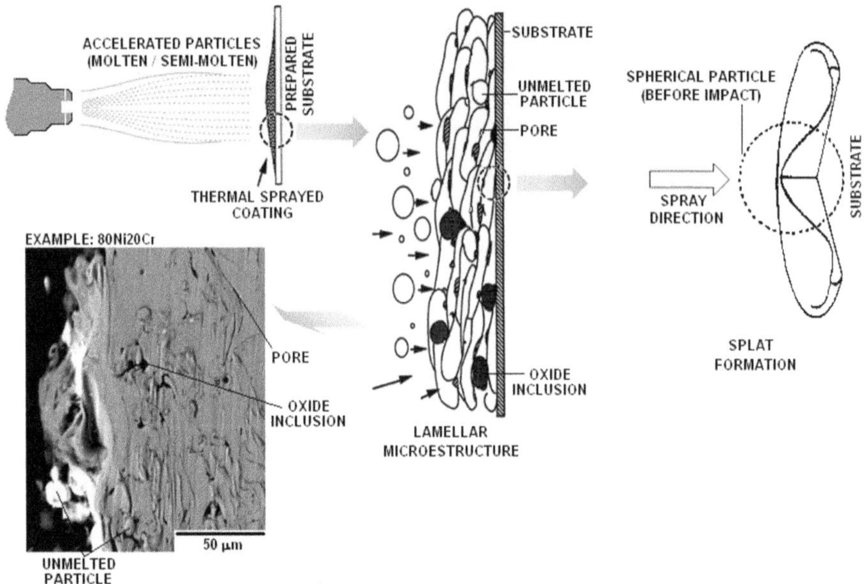

Fig. 11 Schematic diagram of typical microstructure of coatings obtained by thermal spraying.

The homogeneity of the particle properties provides better control in properties of the coating produced by thermal spraying. The particle size distribution and its size, play an important role in this context. Duringtheir course through the flame, small particles are released at the edge. In the area of lower temperature, large particles cannot reach a high plasticity condition. Thus, these particles do not spread adequately on the substrate, which promotes the increase of porosity. The coatings obtained by conventional thermal spray had porosity ranging from 0.025% to 50% (LIMA, 1995). According to Gordon England (2011) the porosity is caused by low impact energy (unmelted particles/low velocity), shadowing effects (unmelted particles / spray angle), shrinkage and stress relieve effects. The porosity can be detrimental in coatings with respect to: corrosion - (sealing of coatings advised), strength, macrohardness and wear characteristics. Moreover porosity can be important with respect to: lubrication (porosity acts as reservoir for lubricants), increasing thermal barrier properties, reducing stress levels and increasing thickness limitations, increasing shock resisting properties.

According to Ctibor and Lechnerova (2006) there are 3 different types of porosity in thermal sprayed coatings. The intrasplat cracks result from the relaxation of residual stresses that developed due to constrained shrinkage of a solidifying splat. The interlamellar pores result from particle fragmentation that causes poor wetting/adhesion between the splats. The globular pores result from a lack of filling around the undulation of the splats as well as from the existence of a missing core due to incomplete melting of the particle in the flame.

The high temperatures in the substrate increase the diffusion between the layers of particles, but also enhance the oxidation of the substrate, which may decrease the cohesion among the particles forming lamellae. For Niemi *et al.* (1995), the high velocity of the particle added to a small grain size produce thinner microstructures and tighter lamellae with better bonds between them. Moreover, high cooling rates or super cooling (10^6 Ks^{-1}) of particles can cause the formation of unusual amorphous (glassy metals) microcrystalline and metastable phases (GORDON ENGLAND, 2011).

Metallic coatings can suffer oxidation during thermal spraying, and the products of oxidation remain usually included in the coating (see Figure 11). Oxides are generally much harder than metal, causing hardening in the coating and higher wear resistant. Furthermore, oxides in coatings can be detrimental towards corrosion, strength and mechanical properties.

In general, the microstructure of a thermally sprayed coating depends on: generation of energy, interaction between energy and spray materials and interaction between spray particles and substrate. Figure 12 summarizes these factors.

For example, Ctibor and Lechnerova (2006) argue that in plasma spray coatings the microstructure development of the coating is governed by interaction between spray droplet and the substrate. The intrinsic properties of the individual splats and the correlation among them are affected by the inflight feedstock-particle properties within the plasma (i.e., their temperature, velocity, size, degree of melting, and extent of particle-particle interaction) as well as the substrate conditions (i.e., its degree of wetting, thermal contact resistance, roughness profile, chemical interactions, and extent of modification by prior deposition of splats).

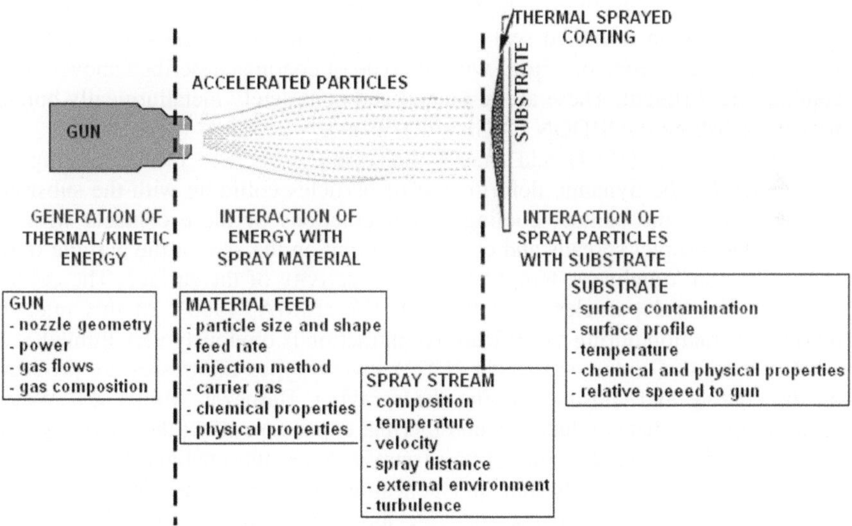

Fig. 12 Schematic diagram of thermal spray factors and the microstructure obtained.

In the interaction between spray particles and substrate, probably the main factor of influence is the adhesion between the substrate and deposited material, that can result in high qualities of the thermal sprayed coatings. Three main factors contribute to adherence of a coating: variation in the spraying distance, pre-heating of the substrate and the use of coatings intermediates, in order to reduce the residual thermal stresses caused by materials with different coefficients of thermal expansion (VARDELLE *et al.*, 1994).

The bonding mechanisms at the thermal spray coating/substrate interface and between the particles making up the thermal spray coating can be divide into three mechanisms: i) mechanical keying or interlocking; diffusion bonding or metallurgical bonding, and other adhesive, chemical and physical bonding mechanisms -oxide films, Van der Waals forces etc. (GORDON ENGLAND, 2011).

Some factors affecting bonding and subsequent build up of the coating are cleanliness, surface area, surface topography or profile, temperature (thermal energy), time (reaction rates & cooling rates etc..), velocity (kinetic energy), physical and chemical properties, physical and chemical reactions.

To obtain adhesion onto the substrate a proper cleaning of the surface is needed, which is achieved by abrasive blasting systems, allowing roughness on the substrate, leading the mechanical anchorage of particles on impact. Moreover, the adhesion depends on the mechanical properties of the substrate, thus factors such as pre-heating and cooling of the substrate after spraying also exert influence.

Increase in thermal and kinetic energy increases chances of metallurgical bonding (temperature, velocity, enthalpy, mass, density and specific heat content etc..). Therefore, higher preheat temperatures for the substrate will increase diffusion bonding activities but will also increase oxidation of the substrate which could defeat the objective of higher bond strengths. Metallurgical or diffusion bonding occurs on a limited scale and to a very limited thickness (0.5 μm max. with heat effected zone of 25μm) with the type of coatings described above. Fused coatings are different. These are re-melted and completely metallurgically bonded with the substrate (GORDON ENGLAND, 2011).

Vardelle *et al.* (1994) add that the physical properties of the coating are determined by the dynamic deformation of particles colliding with the substrate, the reaction of the contact cooling and interactions of these contacts with the surface. The droplet spreads and cools depending on the size of the particle in the state of fusion, roughness, temperature and reactivity of the surface. The cooling rate of individual particles is around 10^6 K/s (LIMA, 1995). At this rate, the thermal interaction during solidification/contraction is obviously very limited.

According to Gerden and Hecht (1972) several types of stresses are produced on the coating and substrate during the spraying. Barbezat, Müller and Walser (1988), reported that conductivity and thermal expansion affect the cooling rate of sprayed particles and the conditions of stress inside the coating. These stresses lead to a decrease in cohesion and adhesion of the coating. The temperature gradient produced between the substrate and the coating is probably the main cause of the residual stresses: i) at the substrate/coating interface layer, potentially causing microcracks in the coating, ii) increasing layer thickness, and iii) during cooling of the already coated substrate, macrocracks can be present, or the detachment of the coating itself may occur. Several properties in the coating layers can be influenced by the nature of the residual stress, adherence, resistance to thermal shock, thermal fatigue resistance and wear resistance.

Particles strike in the substrate and rapidly cool and solidify, forming a lamella. This process generates a tensile stress within the particle and a compressive stress within the surface of the substrate (Figure 13). As the coating is built up, so are the tensile stresses in the coating. With a lot of coatings a thickness will be reached where the tensile stresses will exceed that of the bond strength or cohesive strength and coating failure will occur (GORDON ENGLAND, 2011).

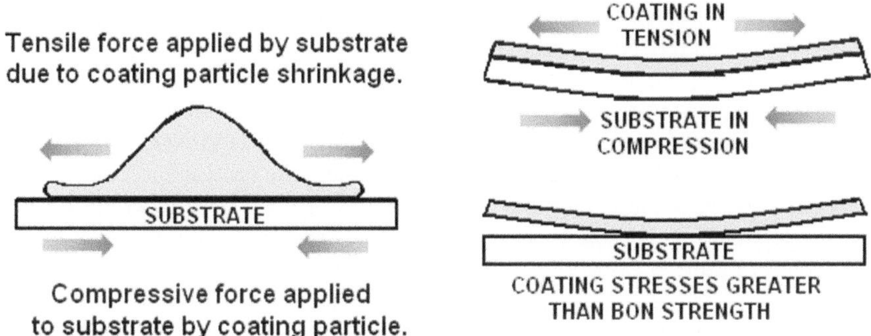

Fig. 13 Schematic diagram of the stress between substrate and coating due to the thermal spraying process. (GORDON ENGLAND, 2011).

Spraying method and coating microstructure influence the level of stress build up in coatings. Dense coatings are generally more stressed than porous coatings. Notice that Combustion powder sprayed coatings generally have greater thickness limitations than plasma coatings. On the other hand the systems using very high kinetic energy and low thermal energy (HVOF, HEP, cold spray) can produce relatively stress free coatings that are extremely dense. This is thought to be due to compressive stresses formed from mechanical deformation (similar to shot peening) during particle impact counteracting the tensile shrinkage stresses caused by solidification and cooling (GORDON ENGLAND, 2011).

3 Erosion

Wear is defined as the progressive loss of material from the surface due to mechanical factors, that is, by contact and relative dislocation with another solid, liquid, or gas (HOPPERT, 1989). It is a valid assumption that, in some situations, chemical phenomenon, such as oxidation, coexists, providing even more drastic conditions for degradation of the material in use.

According to ASTM G 40-92, the erosion is the progressive loss of material from a solid surface due to the mechanical interaction between the surface and some fluid, multicomponent fluid or liquid or solid impact particles.

Hoppert (1989) suggests that the erosive wear occurs when the materials are carried not parallel to surface flow, but when they fall over the surface due to gravity or centrifugal forces. This fall damages the surface, leading to material loss by fracture.

According to Finnie (1995), the phenomenon of erosion has been studied since the 19th century, but the first technical article about erosion appeared in the beginning of the 20th century. This means that the phenomenon of erosion in materials has been observed long time ago in many technological and engineering systems. However, a more reasoned analysis about the erosion process has been performed later, considering the analysis of the structures of the eroded surfaces (SHEWMON and SUNDARARAJAN, 1983).

Many models and mechanisms about the loss of material in the erosion process were already proposed, and they attempted to relate the erosion rate to physical and mechanical properties of the material (HUTCHINGS, 1981 and HUSSAINOVA *et al.*, 2001). In order to understand the erosion mechanisms, it is necessary to know the nature and the magnitude of the forces actuating between the erodent particle and the surface of the target material during the brief contact. These forces transfer energy from the particle to the target material and determine the extension and the morphology of the resultant strain caused by the impact, which can lead to a removal from the impacted material. The interaction between particle and target material suggests that ductile and brittle materials show different responses to erosive wear.

According to Berthier (1989), in ductile materials, such as metals and polymers, the erosive wear occurs preferentially through plastic deformation, by decrusting or cutting, from the surface. Whereas brittle materials, such as ceramics, are highly susceptible to cracks and microfissures. Therefore, they are removed by the interconnection of cracks, which migrate from the point of attack of the particle on the surface (HOPPERT, 1990).

Regarding the method for the quantification of the phenomenon of erosive wear, it can be expressed by the loss of surface mass, related, or not, to the mass of the impacting particles. In some cases, a mass gain can occur, due to incrustation of impacting particles, or even due to oxidation of the surface when the temperature is high enough (FINNIE, 1995).

The response of the different materials to the erosive wear also depends on the conditions during which this wear occurs because some variables affect the extension of the erosion. According to Ball (1986), the main variables are, among others, size, nature and mass of erodent particle, type and velocity of flow, and impact angle of the erodent particles.

Finnie (1995) describes that the resultant wear caused by the impact of a flow of particles over the surface of a material depends on factors such as conditions of the incident flow over the material surface and properties of the material and incident particles. The main factors responsible for the erosion process can be classified as:

i) **Operational:** particle velocity, attack angle, temperature, number of particles per surface unit per time unit, medium corrosivity;
ii) **Properties of erodent particles:** kind of material, size, shape, physical and mechanical properties;
iii) **Surface properties:** kind of material, morphology, tension level, physical and mechanical properties, rugosity, grain size, and porosity.

As already mentioned, the erosion in ductile materials occurs in a different manner similar to that in brittle materials, so a small review of the possible phenomena related to each type of wear will be shown in the following text.

3.1 Erosion in Metallic Materials

In order to describe the mechanisms responsible for erosive wear in metallic materials, it is necessary to know the contribution of certain factors (force, stress, strain) to the impact of a solid particle on the surface of a material, as well as its response to the wear phenomenon.

When a particle strikes a metal surface, the response of this material can be an elastic or plastic deformation, depending mainly on its yield strength. According to Hutchings (1979), one way to know the extension of a damage caused by the impact of a particle on a ductile material can be estimated using the Best or Metz Number (B), given by the Equation 1.

$$B = \rho \frac{V^2}{\sigma_Y} \tag{1}$$

where:

B = Best or Metz Number (dimensionless);
ρ = Density of the target material (Mg/m^3);
V = particle impact velocity (m/s);
σ_Y = Yield strength of the target material (MPa).

Table 4 shows the deformation type expected for the impact of a particle over a wide range of Best's Numbers.

Table 4 Type of damage as function of the Best's Number. (HUTCHINGS, 1979).

Best's Number	Damage regimen
10^{-5}	elastic, *quasi* static
10^{-3}	beginning of plastic deformation
10^{1}	extensive plastic deformation
10^{3}	hyper-velocity phenomenon

Analyzing the obtained values of Best's Number and evaluating the velocity values found in the most part of applications where the erosive wear is detected, it is observed that the damage caused to a metallic material is, most of the time, under plastic deformation regimen, that is, when the value of Best's Number is between 10^{-3} and 10^{1}.

So, considering only plastic deformation of a metal when its surface is struck by erodent hard particles, three different wear mechanisms can be considered, according to Hutchings (1979), Hutchings (1981a), Hutchings (1981b), Cousens and Hutchings (1983), Hutchings (1989), and Finnie (1995): i) for oblique impacts, the ploughing mechanism; ii) cutting mechanism (types I and II) and iii) for near-normal impacts, the detachment by platelet. According to these authors, the mechanisms of material removal with oblique impact are already well defined, whereas for near-normal impacts, there is no accordance among the mechanisms described in the bibliography.

In the ploughing mechanism, the displaced metal is extruded in a border at the end of the impact crater and, depending on the impact angle and velocity, the border can detach, forming a source of mass loss (HUTCHINGS *et al.*, 1976). The energy balance between the incident particle and the material that is struck is described by Hutchings (1977) and is shown in Figure 14. Figure 14, based on the results of Hutchings *et al.* (1976), indicates that for the impact on the material by ploughing mechanism by a spherical particle at 30° angle, approximately 40% of the initial energy is consumed for the formation of indentation and, then, erosion occurs.

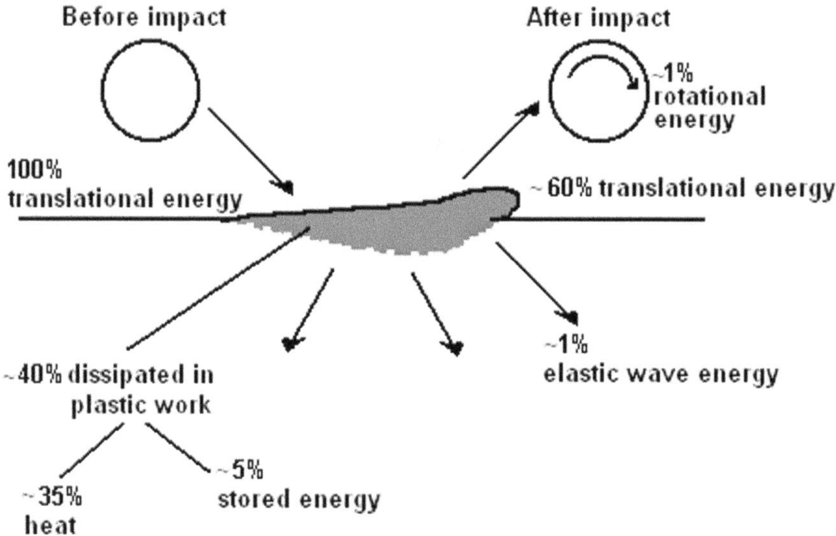

Fig. 14 Energy balance in the impact of spherical erodent particles at impact angle of 30°. The material removal occurs by the ploughing mechanism (HUTCHINGS, 1977).

It is presumed that in the cutting mechanism I, the angular particle rotates forward during the impact, resulting in the indentation of the metal and removal of a big chip of material at the end of the crater. Although this chip generally does not detach in only one impact, it is clearly susceptible to removal by the impact of a succeeding particle that is properly oriented. In this case, the energetic balance between the incident particle and the struck material is more complex, since the particle shape and orientation are important. Figure 15, based on the results of Hutchings (1977), demonstrates that the rotational kinetic energy of the recoiling of the particle can be considered, and it is estimated that 40–80% of the kinetic energy be dissipated during plastic deformation. In the cutting mechanism II, the particle rotates backward during the impact, removing a chip from the material in a way similar to a machining action. The type II cutting is favored only over a narrow range of impact angles and particle orientation and, therefore, occurs less often than type I cutting (HUTCHINGS, 1977).

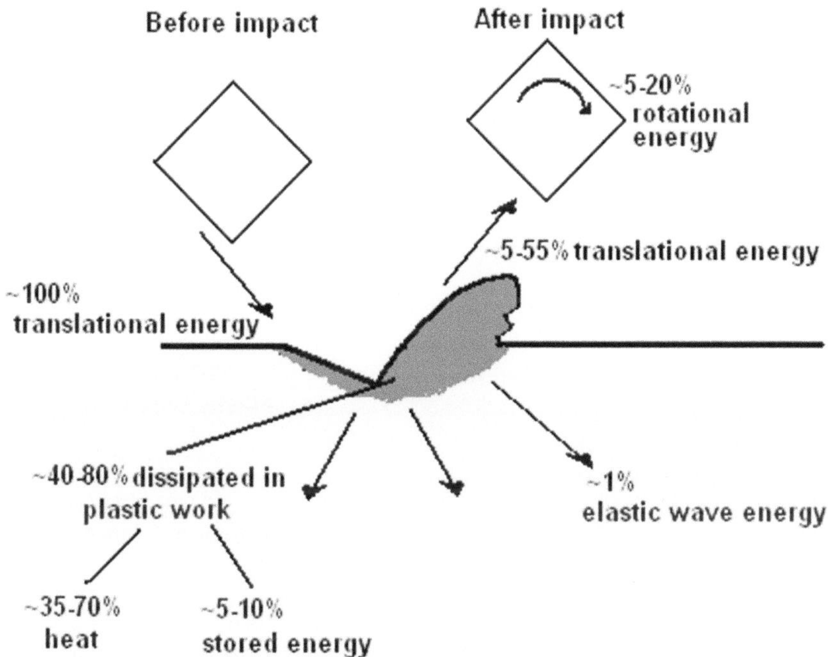

Fig. 15 Energy balance in the impact of angular erodent particles at impact angle of 30°. The material removal occurs by the type I cutting mechanism (HUTCHINGS, 1977).

Hutchings (1989) describes that the mechanism of erosion of metallic materials - for near-normal impacts - can be divided into three distinct phases, which occur sequentially. At the initial phase, the impact of the particle results in the formation of a crater and the material is extruded or detached from this crater. At the second phase, the displaced metal is deformed by subsequent impacts, which can lead to a lateral displacement of the material, which is detached from the surface, or also suffer ductile fracture in the areas with increased deformation. Finally, after relatively few impacts, the detached material become severely deformed, and then it is detached from the surface by ductile fracture. This mechanism has been called *platelet* (LEVY *et al.*, 1984) and is different from the micro-cutting suggested by Finnie (1960), where it is postulated that the impact of several particles is necessary for the removal of a metal fragment from the surface.

Hutchings (1989) emphasizes that this *platelet* mechanism can occur and is observed for any impact angle up to 90°; however, at lower angles (less than 20°) the three phases of the mechanism can occur simultaneously, and it is very difficult to distinguish it from cutting mechanism. An energy balance during the impact of solid particles on the surface of a metal, at an angle of 90°, is shown in Figure 16.

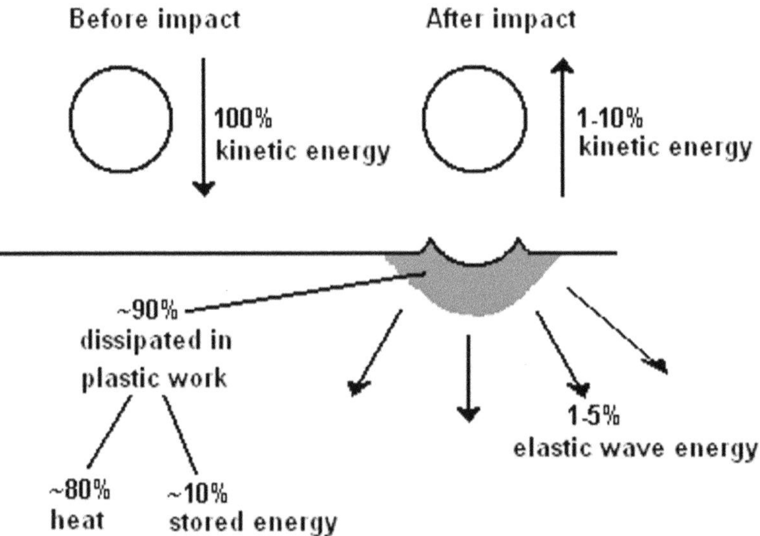

Fig. 16 Energy balance during the impact of erodent spherical particles for impact angle of 90° (HUTCHINGS, 1977).

As shown in Figure 16, 1–10% of the initial kinetic energy of a particle (assumed spherical) in a perpendicular collision is returned to the particle by elastic forces. Much of the initial energy is dissipated on the target. The energy contained in the elastic wave field, which cannot contribute to erosion, can be estimated, and it is described as 1–5% of the initial energy, and remaining approximately 90% is used up in plastic deformation. So, up to 10% of the energy will be stored in discordances and other crystalline imperfections of the metal, and remaining 80% of the kinetic energy of the particle is dissipated as heat. For strongly cold-hardened metals, the fraction of energy stored will be smaller and more energy will be dissipated as heat. In this analysis, a very small contribution of surface energy as well as the kinetic energy of the eroded fragments was ignored.

3.2 Erosion in Ceramic Materials

Ceramic materials show chemical stability, hardness, and mechanical strength at high temperatures. In addition, when compared with metals, they are much less susceptible to damage by corrosive processes. However, because of their brittle nature, the erosion occurs by distinct mechanisms, that is, by propagation and intersection of the cracks produced by the impact of the erodent particles. Madruga *et al.* (1994) schematically suggest the types of flaws and the degradation that occurs in a ceramic material exposed to the attack of solid particles (Figure 17).

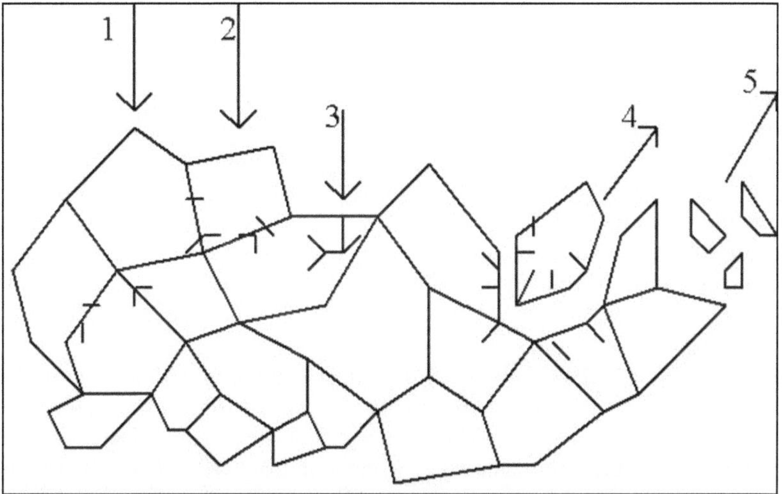

Fig. 17 Wear mechanism. 1. Fatigue in the intergranular phase; 2. Microcracks in the grain boundary; 3.Induction of microcracks in the grain; 4.Detachment of the grain; 5. Detachment of the grain fragments (MADRUGA*et al.*, 1994).

Regarding the erosive wear mechanisms in ceramic materials, Zhou and Bahadur (1993) as well as Butler (1989), Kato (1990), and Bhushan and Sibley (1981) used the mechanism of fracture resulting from indentation which is caused by the impact of spherical particles to explain the erosion of ceramic materials. Basically, two theories have been generally accepted for explaining the mechanism of fracture of these materials. One of them is based on the mechanism of purely elastic fracture and the other on the elasto-plastic fracture mechanism. These mechanisms will depend on the size of the impacting particles.

Purely elastic fracture occurs in particles where the radius at the impact point is greater than the critical radius (greater than 200 μm) and produces conical cracks called *Hertzian*'s cracks. However, it is necessary that an intersection exists among several conical cracks so that the material is detached from the surface (ZHOU and BAHADUR, 1993). The other theory, the elasto-plastic fracture, is applicable to small particles which produce radial and lateral cracks. In this case, the removal of the material can occur without the intersection among cracks. Zhou and

Bahadur (1993) consider two possible behaviors in the elasto-plastic model: quasi-static and dynamic. The greatest difference between them is that in the dynamic behavior, the calculation of the impact force of the particles includes the effect of the dynamic stress and in the quasi-static theory, the kinetic energy of the particles is totally used up in plastic deformation.

There is agreement and discrepancy between the theoretical and experimental results: on the one hand, in glass erosion tests, where large spherical particles strike the surface, some researchers such as Finnie (1960) found clear evidences of material removal from the surface by the intersection of *Hertzian*'s conical cracks. On the other hand, Sheldon (1970) and Sheldon and Finnie (1966) did not get to observe these cracks in their experiments. Ritter *et al.* (1984) and Evans (1982) observed intergranular cleavage and fractures in a pit formed on the surface of an alumina eroded by SiC particles with particle size of 508 μm at a velocity of 75 m/s. However, they did not observe radial cracks in the contact region, as they expected. This type of pit erosion can be formed due to extensive fracture of grain boundaries by the impact of the particles. Similar phenomenon was also observed by Wiederhorn and Hockey (1983) in the erosion of alumina.

Morrison *et al.* (1985) analyzed the impact produced in mullite ($3Al_2O_3. 2SiO_2$) by alumina particles with particle size of 270 μm at velocity of 100 m/s, at impact angle of 90°. In this case, the area that was struck was a central crater, with radial and lateral cracks having dimensions close to those of the impacting particles. Similar observations were made by the same authors in case of an erodent material with average size of 37 μm.

Soderberg *et al.* (1981) observed the imperfections in ceramics with alumina content of 99.7, 99, and 94% caused by the impact of particles at angles of 45° and velocity of 66 m/s. The observed imperfections were basically caused by intergranular fractures. In the case of the ceramic material with 99.7% of alumina, fragments of the erodent material were found. The erosion mechanism was considered to be similar in the three cases.

Zhou and Bahadur (1993) analyzed the impact of particles at angles of 90° on aluminas with different amount of glass and of zirconia, as well as high-purity alumina, at room temperature. They used SiC grain with particle 120 μm and velocity of 50 m/s as the erodent. In this study, the authors observed that the mechanism of fracture was basically intergranular type without traces of plastic deformation. In some cases, cracks were formed and propagated without detachment of grains from the surface. Some radial cracks were observed from the impacted edges of the region. This is typical for a fracture by indentation which leads to lateral crack formation. In the case of zirconia-added alumina, transgranular cracks were observed.

Wensink and Elwenspoek (2002) who studied the impact of solid particles on the surface of a brittle material observed that the contact area between particle and target material is plastically deformed due to high compression; besides that, shearing stresses and a radial crack were formed. After the impact, plastic deformation leads to an increase in the internal stresses of the material, which causes the removal of particles from its surface. Figure 18 shows in a simplified way the formation of a fracture by the so-called lateral fracture mechansim.

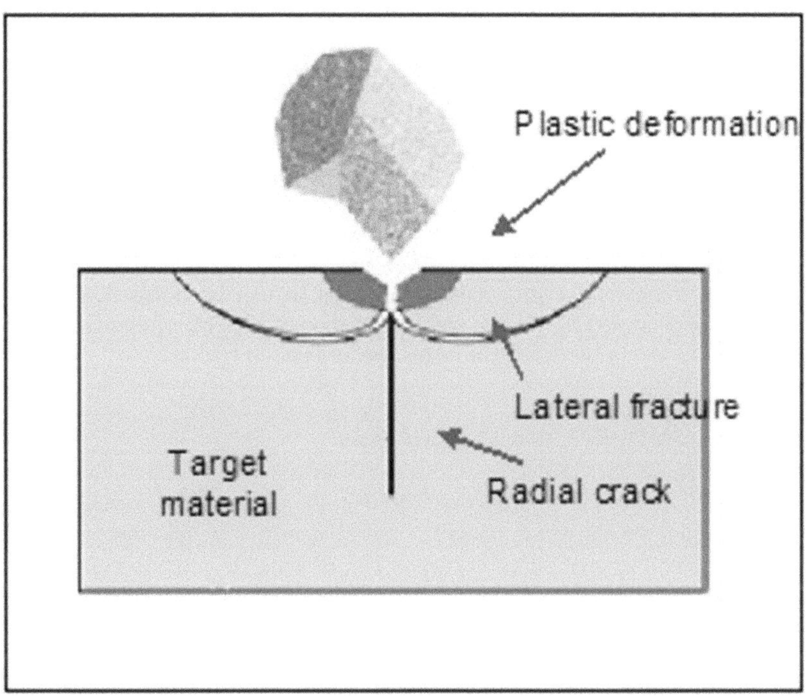

Fig. 18 Formation of a lateral fracture caused by the impact of hard particles on the material surface. The lateral fracture is formed by the release of stresses in the plastically deformed area after generating a radial fracture (WENSINK and ELWENSPOEK, 2002).

Marques (2006) evaluated the erosion strength of alumina-based ceramic bodies, with and without vitreous phase, eroded by electrofused alumina, and observed that the wear rate is fundamentally influenced by the present phases, particularly, porosity (size and amount) and vitreous phase (distribution and amount). In addition, it has been noted that the temperature has decisive influence in increasing the wear of ceramic bodies under erosion. It was evident that the erosion rate depends on the porosity of ceramic bodies made up of only alumina, since the contact points (or fixation) of alumina particles decrease significantly as the porosity increases. As a consequence, the microstructure becomes more susceptible to the damage when erosion occurs.

Comparing ceramic bodies with and without vitreous phases, Marques (2006) observed that the alumina with vitreous phase presented higher erosion rate than the alumina with less porosity, increasing with the temperature and with the impact angle of the incident particles, for all investigated conditions. The vitreous phase offers a target less resistant to wear than the alumina phase. However, the alumina with vitreous phase presented erosion rate significantly lower than porous alumina over the whole range of temperatures and for all the investigated attack angles. This fact was explained by the reinforcement given by the vitreous phase in the porous alumina microstructure, strengthening the bond among alumina particles.

Concerning erosion mechanisms, Marques (2006) noticed that the main mechanism which leads to erosive wear of alumina, with and without vitreous phase, is the formation of erosion pits. This happens mainly at room temperature and at incident angles of 90º. This process is based on brittle fracture of the material and occurs by microcracking throughout the grain boundary of alumina. As the temperature rises, in the presence of vitreous phase, one can notice loss of material also by ductile fracture, due to the plastic deformation of vitreous phase by viscous flow. As a result, a rounding of the edge of the erosion pits occurred, with the increase of temperature and glass content.

From the review on brittle material erosion, it can be noted that two main theories were proposed by several authors for the removal of brittle material under the impact of erodent particles: pit formation and lateral fracture.

4 Erosion in Bulk Cermet Materials

The classical theory of erosion predicts that the response of a material under erosion depends on its nature, ductile or brittle; however, this theory cannot be used to explain the erosion of cermets. The erosion of cermets points that the nature of this degradation is very complex, which is attributed to its nonhomogeneous character.

According to Hussainova (2001) *apud* Hussainova *et al.* (2001), the erosion of bulk WC-Co- and TiC-based cermets is associated with the combination of ductility and brittleness in erosion, although brittleness is dominant. Recent works by Feng and Ball (1999) *apud* Hussainova *et al.* (2001) attempted to define erosion regimens of bulk cermets ranging from plasticity-dominant to fracture-dominant behaviors, but Hussainova *et al.* (2001) suggest that this transition is not well defined.

The study of erosion in cermets obtained by powder metallurgy forms the basis for evaluating the erosion in coating cermets obtained by thermal aspersion technique. According to Hussainova *et al.* (2001), in the erosion process, the fracture of a bulk cermet material begins, generally, in the binding phase (ductile matrix). The carbide grains lose their protective bond and the eroded surface is almost completely covered, with the exposure of the carbides. If the hardness of the target material is greater than the hardness of the erodent, the erodent particles hardly can cause a deformation, and the erosion of the cermet predominates. The elastic penetration degree and, therefore, the amount of energy transmitted to the surface depend on the modulus of elasticity. If it is high, elastic penetration is less. Under these conditions, the impact of erodent particles can cause low-cycle fatigue failure in the matrix and in the carbide grains. So, the modulus of elasticity is a more important parameter to the wear strength than the hardness. However, if the hardness of the erodent material is greater than the hardness of the cermet, processes of penetration of the erosion occurs on the surface of the material, through cutting and ploughing mechanisms, and cracks are formed in the large carbide grains, resulting in the detachment of small chips. Under these conditions, there is no distinction among the different classes of cermets investigated, with metallic matrix content increasing from 8 to 60%, in WC, TiC, or Cr_3C_2 carbides (HUSSAINOVA *et al.*, 2001).

This result for wear strength as a function of the modulus of elasticity and hardness can be observed in Figure 19. In fact, according to Hussainova *et al.* (2001), no linear relationship is observed between hardness and wear strength (Figure 19a). Similar conclusion was given by Reshetnyak and Kübarsepp (1994), when they investigated 16 TiC-Ni-based alloys, in which they noticed that, to the same level of hardness, the wear strength of cermets varied up to 80%. Therefore, for erosive wear strength, the evaluation of hardness is only a first approach.

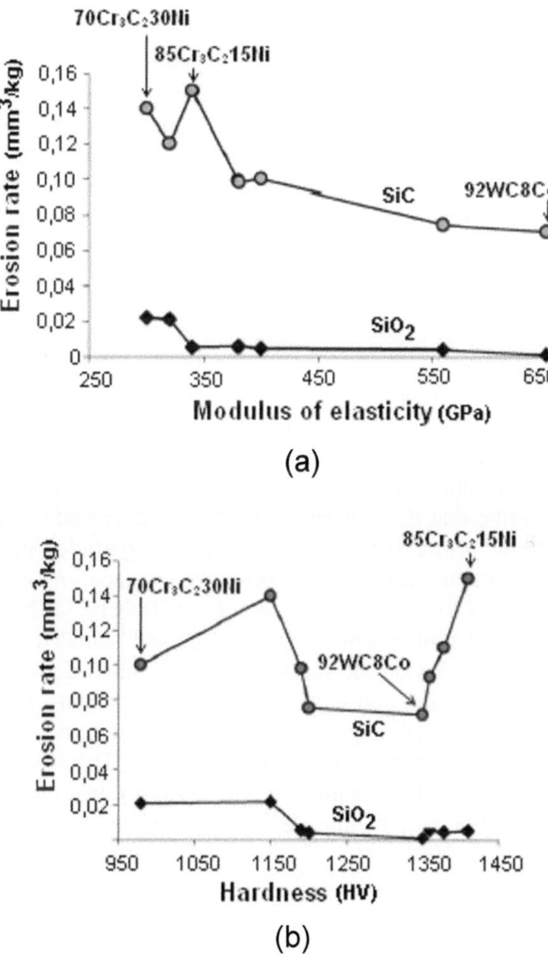

Fig. 19 Erosion rate as function (a) of the hardness and (b) of the modulus of elasticity for different bulk *cermets* eroded by SiO_2 or SiC (HUSSAINOVA *et al.*, 2001).

If the erosion of brittle grains (carbide phase) preferentially occurs by a mechanism which includes the initiation and propagation of microcracks, then the fracture tenacity of the material affects the erosion rate. The low tenacity of many hard components results in a loss of erosion strength, which explains the low performance of cermets based on Cr_3C_2 (fracture tenacity of $3MPa.m^{1/2}$) and the possible beneficial effect on cermets based on WC (fracture tenacity of $15MPa.m^{1/2}$). This means that the WC grains are hard and sufficiently strong to dissipate the large amount of energy caused by the striking of the erodent particles, without the microcracking of the grains. Therefore, the mechanical properties and the wear strength of cermets depend on each individual grain, the matrix, and the bonding among phases and the adjacent grains (WELLMAN and ALLEN, 1995 and ENGQVIST *et al.*, 1999 *apud* HUSSAINOVA *et al.*, 2001).

Hussainova *et al.* (2001) evaluated the behavior of erosive wear of Cr_3C_2-based cermets and noticed that if the erodent is much harder than the target material (in this case, SiC), the initiation of the cracks is inevitable and the crack propagation rate is the factor which controls the erosion. The amount of brittle phase (carbide grains) of 85 wt.% results in poor wear strength, which can be attributed to its low fracture tenacity and weak bonding among the grains and the metallic matrix. In the case in which the hardness of the erodent is less than the hardness of the target material (in this case, SiO_2), for the same cermet with 85 wt.% of carbides ($85Cr_3C_215Ni$), there is an optimal wear strength, which can be attributed to its high plastic penetration strength. Figures 20 and 21 show micrographs of $85Cr_3C_215Ni$ bulk cermet after erosion at angle of 45° by erodent particles of SiC and SiO_2, respectively.

Analyzing the micrographs of the eroded surfaces (Figures 20 and 21), it is evident that the erodent particles with greater hardness (SiC – Figure 20a) cause considerable damage on the surface of the tested cermet ($85Cr_3C_215Ni$). Observing, in Figure 20b, the cross-section of cermet at eroded area, Hussainova *et al.* (2001) describe that the material removal probably had occurred due to the formation of lateral brittle fracture. The impact of particles generates residual stresses, causing lateral cracks just below the surface, making the material removal by decrusting easier. However, the cermet ($85Cr_3C_215Ni$) surface eroded by SiO_2 shows a different morphology, smooth and rounded (Figure 21). The impact of erodent particles on the surface results in small detachment of carbide grains in each collision and consequent gradual detachment of binding material (metallic matrix). Small carbide grains lose their protective bonding and collapse, whereas the large grains are fragmented due to fatigue. The erosion caused by the mechanism of lateral crack formation in SiC leads to erosion rates higher than those in the erosion caused by the decrusting mechanisms that occur in SiO_2 (HUSSAINOVA *et al.*, 2001).

Hussainova *et al.* (2001) observed, besides the microstructure, the effect of impact angle and the velocity of the particles on the erosion strength of three different classes of cermets, namely, WC-, TiC-, and Cr_3C_2-based cermets. The authors noticed that with the decrease of the impact angle, the advantages of WC over TiC cermets decrease. At near-normal angles, the difference between the erosion rates of these two classes become more significant. Nevertheless, decrease in the erodent velocity results in a decrease in the difference in erosion strength between cermets based on WC and those based on TiC.

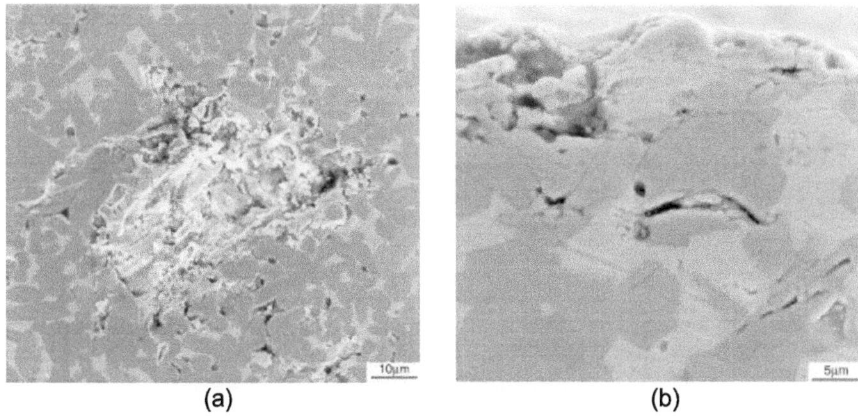

(a)   (b)

Fig. 20 SEM micrographs (a) of a single crater in $85Cr_3C_215Ni$ impacted by SiC particles at attack angle of 45° and velocity of 31 m.s^{-1} and (b) cross-section of the eroded cermet surface (HUSSAINOVA *et al.*, 2001).

Reshetnyak and Kübarsepp (1994) and Larsen-Basse (1983) emphasize that the erosion mechanism depends, first of all, on the test conditions. However, according to Hussainova *et al.* (2001), regardless of the hardness of the erodent material at low impact velocities, the erodent material can cause a local strong destruction of carbide grains and their bonding with the matrix is damaged, this being the weakest phase prevailing, and then selective erosion occurs. Under high-velocity conditions, failure can occur by low cycle fatigue of the carbide grains and by microcutting, depending on the ratio between the target and the hardness of the erodent material.

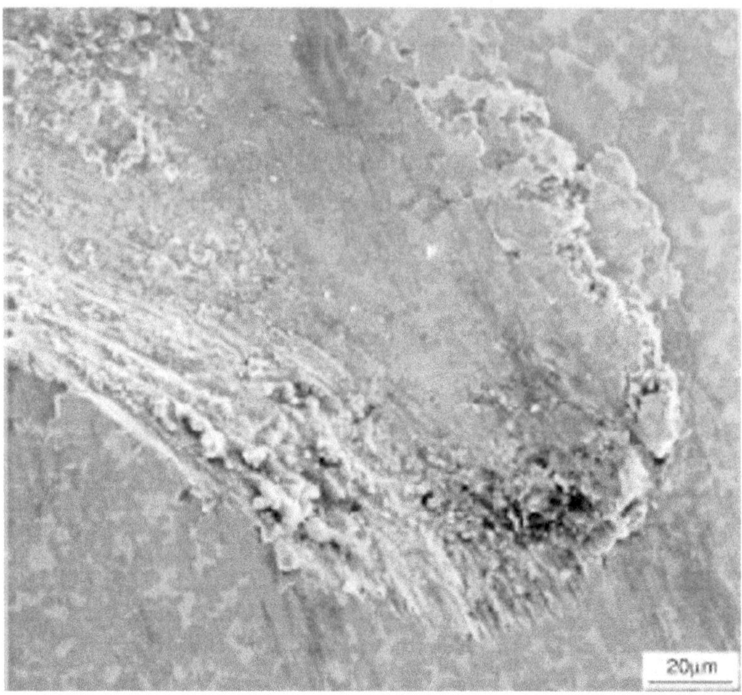

Fig. 21 SEM micrograph of a single crater in $85Cr_3C_215Ni$ impacted by SiO_2 particles at attack angle of 45° and velocity of $31m.s^{-1}$ (HUSSAINOVA *et al.*, 2001).

The wear mechanism of cermets can cause the extraction of the metallic matrix, since the successive impact of erodent particles on the surface of cermet leads to the formation of stress on the surface and the carbide grains interact with each other, expelling part of the matrix. The partial removal of the matrix originates in stress-concentrated areas, which leads to the nucleation of cracks, which initiates the ductile fracture of the material (HUSSAINOVA *et al.*, 2001).

Bergman *et al.* (1997) investigate the influence of the content of carbide and the presence of differently sized carbides in high-speed steels obtained by powder metallurgy. The authors defined primary carbides as carbides with dimensions greater than 1–10 μm and secondary carbides as those with dimensions lesser than 100 nm. According to the authors, the presence of carbides decreases the erosion rate of the material eroded by SiO_2 (considered as erodent with low hardness). The presence of larger carbides did not modify this tendency, since the mechanism responsible for erosive wear is based on cumulative indentations, resulting in micro-fatigue and consequent loss of material. The amount of larger carbide particles in steel is more important than the carbide size, which is attributed to the fact that the small carbides are responsible for the prevention of erosion by the impact of particles through penetration. However, the use of erodent materials harder than the larger carbides cause the preferential erosion of the ductile matrix, indicating that the presence of these carbides in high-speed steel does not improve the erosion strength of the material.

The main erosive wear mechanisms of cermets are still an open issue. Wayne *et al.* (1989) adapted the model of Evans *et al.* (1976) of brittle response of surfaces caused by impact of hard particles. In this case, the wear is a consequence of the lack of capacity of the material to resist the sequence of effects (indentation by the particles, subsuperficial development of lateral cracks, propagation to the surface, micro-delamination, and material removal) before deterioration. The fitting of this model to cermets relates wear strength (W^1), hardness (H), fracture tenacity (K_{IC}), and microstructural parameters such as average size of carbide particles (G) and average thickness of intergranular ductile phase (λ), as shown in Equation 2.

$$W^{-1} \propto \frac{(K_{IC} \cdot H^{\frac{1}{2}} \cdot \lambda)}{G} \tag{2}$$

where

W^1 = erosion strength (kg/mm^3);
G = average size of carbide particles (mm);
λ = average thickness of the intergranular ductile phase (mm).

The average thickness, λ, in turn, can be related to the volumetric fraction (f_V) of the metallic binding phase according to Equation 3.

$$\lambda = \frac{(f_V \cdot G)}{(1 - f_V)} \tag{3}$$

In more recent works, Yao *et al.* (2000) and Beste *et al.* (2001) demonstrate that the dependency of erosion strength on the structural parameters deviates from the model of brittle wear when carbide powders with ultrafine particle size (0.6 μm) are used. Higher contents of matrix result in higher wear rates than those observed for lower contents of this phase, which is attributed to the fact that the significant increase of the average path among the carbide particles (λ) emphasizes the ductile response mechanism of the composite phase of the erosive action. This factor is responsible for the low content of binding phase (ductile matrix) of the ultrafine grains of cermets, sometimes with drastic reduction to ≤ 1 wt.%. Utilization of ultrafine particles also facilitates the uniform and continuous distribution of the metallic phase among the carbide grains.

5 Erosion in Cermet Coatings

Correlation of the different classes of materials that are subjected to erosive wear gives rise to some strong distinctions. For example, with regards to the impact angle, there is a rule that for low angles, the ceramic materials can be adequately used for reducing the erosive wear due to their high hardness and tenacity properties (FINNIE, 1995). However, if a material is under the impact of particles at a near-normal angle (90°), the exposed surface should be able to resist recurring deformations. In this case, more plastic materials, such as metals, could be preferred over ceramic materials and ceramic coatings, in which the cracks quickly progress and lead to material removal. This relationship between properties of materials and wear strength is shown in Figure 22.

In some situations, there is a combination of mechanisms leading to erosive wear, such as cutting, decrusting, fatigue, and brittle fracture mechanisms. In these cases, there is a compromise between hardness and fracture tenacity of the materials and coatings. According to Kulu et al. (2005), composite coatings of metallic matrix reinforced with ceramic particles partially solve this problem, as presented in Figure 23. The authors highlight that coatings with greater hardness and lesser tenacity are more resistant to the wear at attack angles of erodent particles of up to 30º. The use of more tenacious and less hard coatings is recommended when the incident angles are greater than 60º.

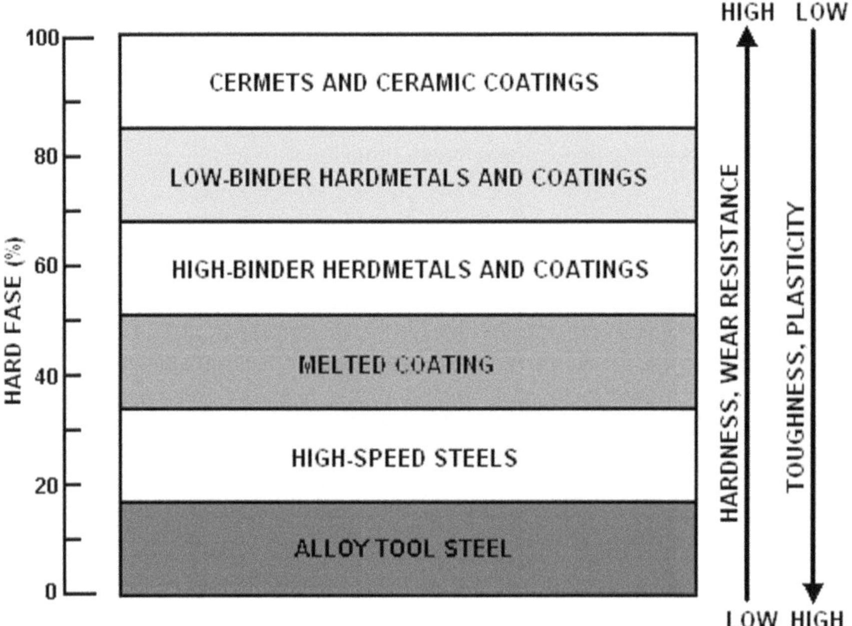

Fig. 22 Wear strength of materials and coatings (KULU *et al.*, 2005).

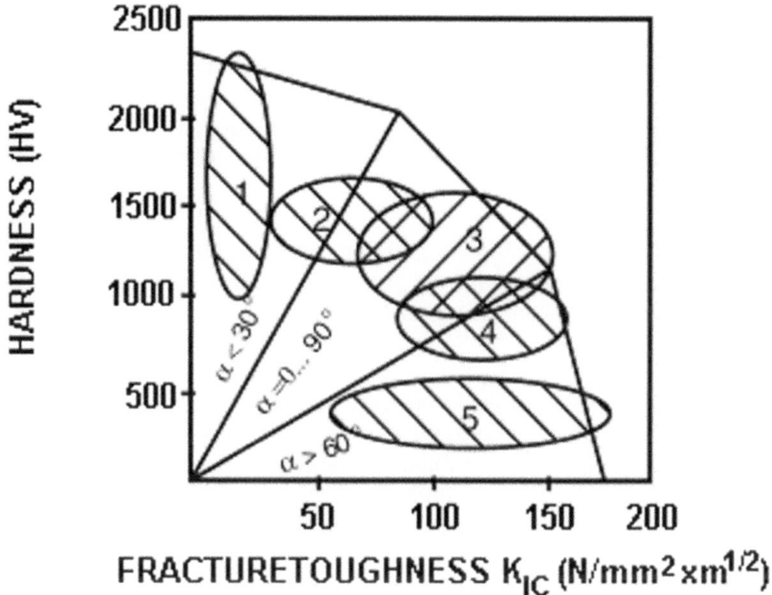

Fig. 23 Hardness and tenacity properties of coatings and their application areas: 1, 2 – low impact angles, 2 a 4 – abrasion and erosion mixture and 4, 5 – high impact angles (KULU *et al.*, 2005).

According to Kulu *et al.* (1998) *apud* Kulu *et al.* (2005), under the wear conditions, when there is a combination of mechanisms, the coatings obtained by thermal aspersion technique based on WC-Co are highly effective. In addition, the materials obtained by thermal aspersion, such as cermets, are often used to resist different wear types, in several industrial applications and also under conditions of extreme erosion.

Wang and Verstak (1999) state that the wide use of cermets as coating materials can be attributed to their combination of positive properties, such as high hardness, mechanical strength, rigidity, and wear strength. However, it is necessary to test different coatings under distinct conditions of wear phenomenon because there is a lack of information about the behavior of different types and classes of cermets.

Tungsten carbides (WC), bonded with Co, are widely used because of its excellent wear strength; however, at high temperatures and in aggressive media, the low corrosion strength of this coating system restricts its application. In these cases, binding, such as iron and nickel, is adopted for the metallic matrix. The more known free tungsten cermets are based on TiC and Cr_3C_2 and they are incorporated into metallic matrices of Ni and Mo alloys.

It should also be observed, according Kulu *et al.* (2005) and Wang and Verstak (1999), that the capacity of the coating that protects the base material (substrate) against erosion depends not only on the composition and microstructure of the coating but also on its mechanical and physical properties, which is in turn dependent on the raw material and its processing.

The advantage of using cermets over using ductile or brittle materials can be related to a combination of physical and mechanical properties of ceramics and metals, the high melting point of the carbides, consequently of the cermets and, therefore, the high thermal and chemical stability of the carbides (UPADHYAYA, 2001).

With the development of application techniques such as HVOF, it became possible to fulfill the increasing needs for protection against erosion, which is due to the improved adherence with the substrate, diminished porosity, and oxide content along with the compressive residual stresses in the coatings (LILLE *et al.*, 2002 *apud* KULU *et al.*, 2005). Wang and Verstak (1999), Wang and Lee (1997), Barbezat *et al.* (1993), and Kulu and Pihl (2002) also state that the evolution of HVOF processes has resulted in the generation of coatings with less decomposition of carbides compared with other deposition processes, due to a combination of high kinetic energy and low aspersion temperatures.

Other authors (KULU *et al.*, 1998 and KULU and VEINTHAL, 2000 *apud* KULU *et al.*, 2005) state that the deposition of coatings by detonation is also promising, due to the excellent wear strength.

The self-fluxing alloys with WC particles applied by the aspersion and fusion techniques (flame, plasma, laser, fusion, etc.) have excellent wear strength and therefore are cost-effective and efficient. Metallic alloys based on MCrBSi, where M can be Ni, Co, or Fe, can be melted at 1050°C. Due to the low porosity and high adherence with the substrate, these alloys can significantly resist impact loads. As coatings, they are used where wear and corrosion strengths are necessary (KULU *et al.*, 2005).

According to Raak (1988) *apud* Levy and Wang (1988), coating microstructures that have high-temperature erosion strength are composed of lamellas and small-sized grains, and have low porosity, less oxide content, good adherence, and no cracks. Hardness, chemical composition, and second phase distribution (carbide particles) have less effect on the wear strength. Moreover, different forms of erodent and erosion test conditions lead to change in the strength of the tested coatings (LEVY and WANG, 1988).

Certain characteristics of coatings (porosity, incorporation of hard phases, content of hard phase, microstructure of formed phases – size of hard phase, form and distribution – hardness and temperature) and the interaction of these with the erosive wear strength are given below.

5.1 *Porosity of the Coatings*

With regards to the porosity of the coatings, Kulu (1989), Kulu and Zimakov (2000a), Kulu and Zimakov (2000b), and Kulu and Veinthal (2000), *apud* Kulu and Pihl (2002) based on the study of erosion strength of 15 coatings sprayed by different techniques (aspersion by flame, plasma spray, D-gun, and HVOF) observed that only the low-porosity coatings (less than 5%) present erosive wear strength. The erosion wear strength of coatings with porosity greater than 5%, even with similar hardness, obtained by different techniques of thermal aspersion, can differ by more than one order of magnitude for identical wear conditions. Therefore, only coatings with low porosity present considerable erosive wear strength.

Levy and Wang (1988) investigated 12 different cermet coatings and they related their erosion strength to the wear mechanism of each coating, where the grain size, the porosity, and the presence of cracks formed during the deposition were considered significant factors.

Cr_3C_2 coated by D-gun technique presented the best performance at 30° than at 90°. Analyzing the coating microstructure, after the aspersion and after erosion at both studied angles, the authors drew some conclusions. The sprayed coating consists of particles with well-defined grains, with a small intergranular porosity, and rugose surface. Regarding the erosion at 30°, the eroded surface has a smooth microstructure (almost similar to a polished structure) and some incrustations of particles. At 90°, the eroded surface is more rugose than that at 30°. The authors consider that the erosion at 90° occurred by cracking and decrusting of small pieces and that their sizes were determined by the basic grain size of the coating. Levy *et al.* (1983) *apud* Levy and Wang (1988) had already concluded that in brittle materials the finest grain size and the less porosity lowers the erosion rate because the decrusting occurs for each grain.

The WC-Co sprayed by plasma spray is a coating which presents a fine network of pores. Levy *et al* (1983) *apud* Levy and Wang (1988) observed that this uniform pore distribution, which is almost equiaxial, lead to deterioration of the erosion strength more than flat pores between lamellas, typical of a coating deposited by flame. This is attributed to the fact that this uniform and equiaxial distribution of pores would expose the material of the matrix with less erosion strength. The morphology of eroded surface of WC-Co at 30° suggests that the remains of the material of the erosion process spread over the eroded material and fill the voids. The size and geometry of pores are directly related to the erosion rate of the sprayed coating. With regards to the Cr_3C_2 deposited by D-gun, it was observed that the less number of pores and the finer form are responsible for a lower erosion rate over WC-Co sprayed by plasma spray. For coatings without a network of cracks, lower porosity and smaller pore size lead to lower erosion rates.

Table 5 Variation of type of erosion rate at impact angles of 30° and 90° for three different coatings (LEVY and WANG, 1988).

Coating	Application Technique	Erosion rate (cm^3/g x 10^{-4})	
		30°	**90°**
Cr_3C_2	*D-gun*	0.063	0.13
WC-Co	Plasma *spray*	0.21	0.56
NiCrBC	Plasma *spray*	0.87	0.61

The NiCrBC sprayed by plasma spray has pores with several sizes, besides disseminated cracks, which makes easier the loss of material and degrades the erosion strength of the coating (LEVY and WANG, 1988). The difference in pore size between two coatings applied by plasma spray (WC-Co and NiCrBC) has a considerable effect on the erosion rate, according to Table 5.

At impact angle of 30°, the NiCrBC coating shows poorer performance than that found at 90°. Therefore, as shown in Table 5, the largest difference in erosion rates occurs at impact angles of 30° and 90° and there is no remarkable difference among these rates (LEVY and WANG, 1988). The authors do not consider the erosion mechanism; however, it should be observed that the NiCrBC coating is metallic, so it is less resistant to erosion at low impact angles.

According to Kulu and Pihl (2002), only coatings with porosity less than 5% are useful as protective coatings against erosive wear. Vicenzi (2007) observed a strong exponential tendency of the wear rate of coatings based on $NiCr-Cr_3C_2$ with the increase in the porosity. Figure 24 presents the results of this research, which indicate that the increase in erosion rate with the increase in porosity occurs in an exponential way, drawing attention to the coatings obtained by HVOF aspersion (less porosity, below 4% shown in detail in Figure 24 - $NiCr_{0,5\%}$, $NiCr35CrC_{3\%}$, $NiCr70CrC_{3\%}$, $NiCr75CrC_{1,5\%}$ - the subscripts are the porosity of each coating) over the coatings obtained by plasma spray (greater porosity - $NiCr_{9\%}$, $NiCr35CrC_{29\%}$, $NiCr70CrC_{31\%}$, $CrC_{28\%}$ - the subscripts are the porosity of each coating).

In coatings obtained by plasma spray, it can be observed that the increase of ceramic phase, for example, 35% of Cr_3C_2 in the NiCr matrix ($NiCr35CrC_{29\%}$), increases the erosion rate, in this case, 14 times at attack angle of 90° and 4 times at 30°, which suggests a strong influence of porosity on the susceptibility of erosive attack. It is possible to suppose that the porosity reduces the contact points between lamellas, making easier the detachment of the material during the erosion. Besides, a change in the erosive mechanism can occur with the increase of ceramic phase, and it appears that the brittle erosion is dominant, due to higher wear rate in the order of attack angles 30°, 45°, 60°, and 90°. A greater increase of Cr_3C_2, that is, 70% in the NiCr matrix ($NiCr70CrC_{31\%}$), leads to the greatest wear rate for the attack angle of 90°. It can be observed that the wear rate increases with the increase of the Cr_3C_2 phase, probably due to which the microstructure becomes brittle because of the greater amount of the ceramic phase.

The author observed a different behavior for coatings with porosity less than 4% (obtained by HVOF). Figure 24 shows the data from HVOF, on a smaller scale, for studying the influence of the porosity. It can be noted that apparently the erosion rate is not significantly influenced by the porosity; in addition, the increase in carbide content in the coating decreases the erosion rate at room temperature. Therefore, for very small porosities, this would not be the fundamental property that controls the erosive wear of the coatings. In fact, the coating with the lowest porosity ($NiCr_{0,5\%}$) presented the highest erosion rate, whereas coatings with similar porosities ($NiCr35CrC_{3\%}$ and $NiCr70CrC_{3\%}$) presented different erosion rates. The erosion rate reaches the lowest values for increments of 70% of Cr_3C_2 in NiCr. It is also noted in Figure 24 shows a change in the attack angle where the highest wear occurs from 30 to 90°, so it is possible that there is an intensification of the brittle character in the erosion of the coatings by an increase in the ceramic phase.

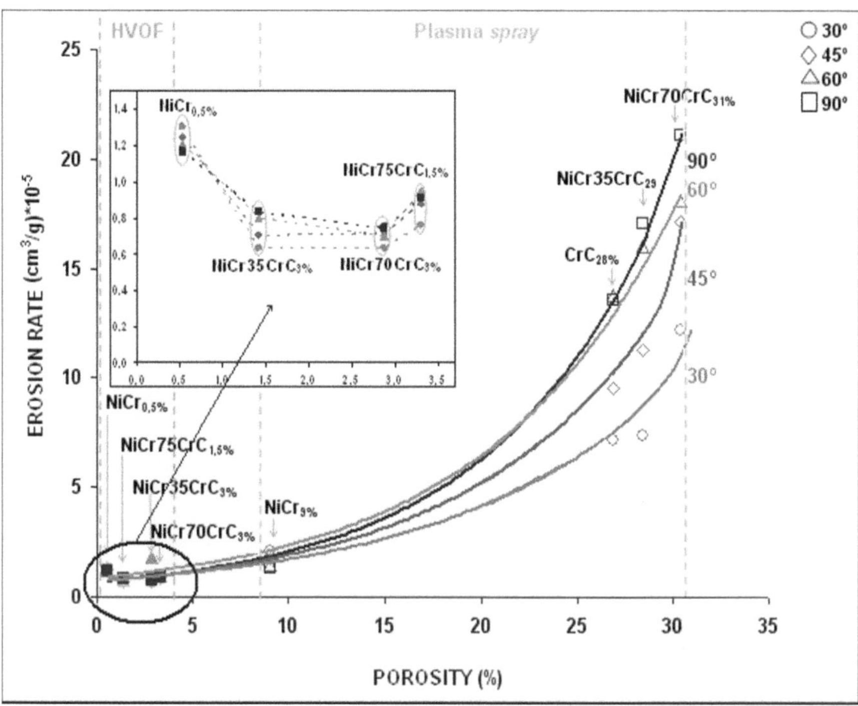

Fig. 24 Erosion rate variation, as loss of volume by mass of the erodent impacted, as function of porosity; normalized by the attack angle of the erodent and by the type of coating (VICENZI, 2007).

Vicenzi (2007) also presented data that point to a differential behavior of the porosity with the increase of temperature, where this property becomes important to protect the coating from erosion and oxidation processes. This fact will be treated in the discussion on high temperatures.

5.2   Incorporation of Hard Phases (Carbides) in the Coatings

Hawthorne *et al.* (1999) related the wear mechanisms responsible for erosion by inserting hard phases in purely metallic coatings. Thermally sprayed coatings based on metals develop mechanisms which are similar to those of ductile bulk materials during degradation by the erosive process. For Ni-based alloys, during erosion at 90°, degradation of the material occurs by cutting and platelet mechanisms (Figure 25). In agreement with the studies of Hawthorne *et al.* (1999), Vicenzi (2007) observed that in pure metallic coatings of NiCr, regardless of its porosity, the wear occurs in the cutting type (for low attack angles, 30°) and in platelet type (for high attack angles, 90°). Figure 26 shows the microstructure of coatings of NiCr with 0.5% of porosity (NiCr₀,₅%) and with 9% of porosity (NiCr₉%).

The erosion strength of cermets depends, besides the attack angle, on the amount of carbide inserted into the matrix. Figure 27 summarizes the results obtained by Hawthorne *et al.* (1999) and Figure 20 the results of Vicenzi (2007), for different types of coatings, with different amounts of hard phases as a function of attack angle.

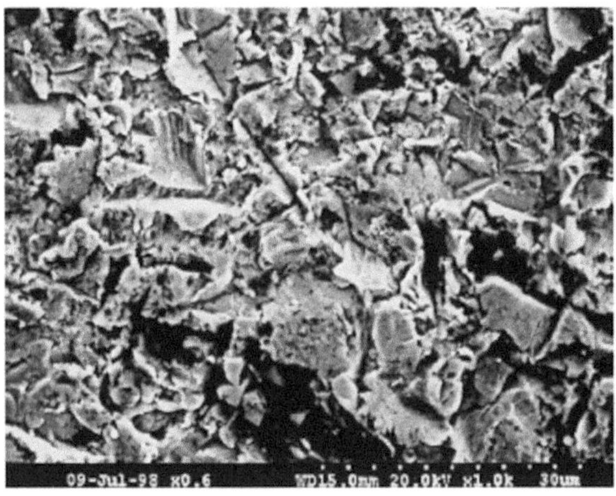

Fig. 25 Typical microstructure of NiWCrSiFeBC coatings after erosion at 90° impact angle (50μm particles of Al_2O_3, velocity of 84 m/s, 25 min. test, 50g of erodent, at room temperature) (HAWTHORNE *et al.*, 1999).

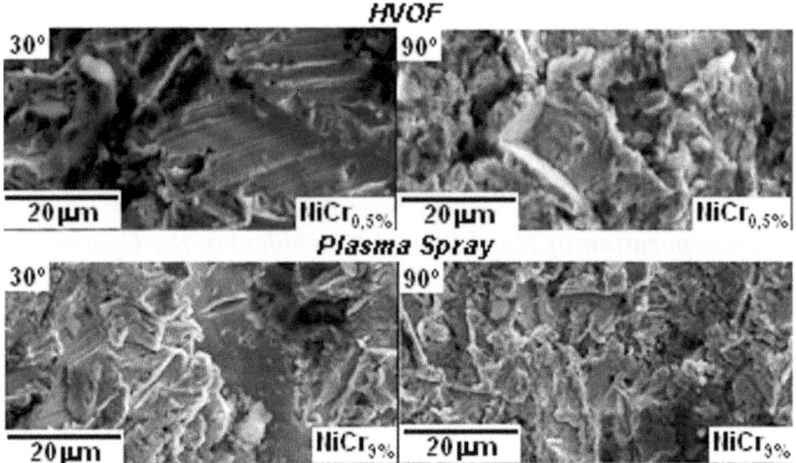

Fig. 26 Typical microstructure of metallic coatings of NiCr after erosion at impact angles of 30° and 90° (180μm particles of electrofused Al_2O_3, velocity of 50 m/s, at room temperature – 10 min test for $NiCr_{9\%}$ and 30 min test for $NiCr_{0,5\%}$) (VICENZI, 2007).

Hawthorne *et al.* (1999) observed that the increase of carbide concentration and the impact angle generally gives higher erosion strength to WC coatings for the three cobalt matrices investigated. However, at low impact angles, the behavior of the three Ni-based coatings was different, that is, opposite to the behavior observed at 90° angle. It is possible that there is certain relevance between the low energy of bonding between the matrix and the Ni particles of WC in the coating obtained by HVOF. Therefore, a higher carbide concentration between 29 and 84% in volume can make the surface more susceptible to the brittle fracture, promoting the detachment of WC by the high impact energy of the erodent. Figure 29 shows evidences of erosion by cutting, formation of platelets, and occasionally by removal of individual carbides after the erosion of the WC-12Co coating.

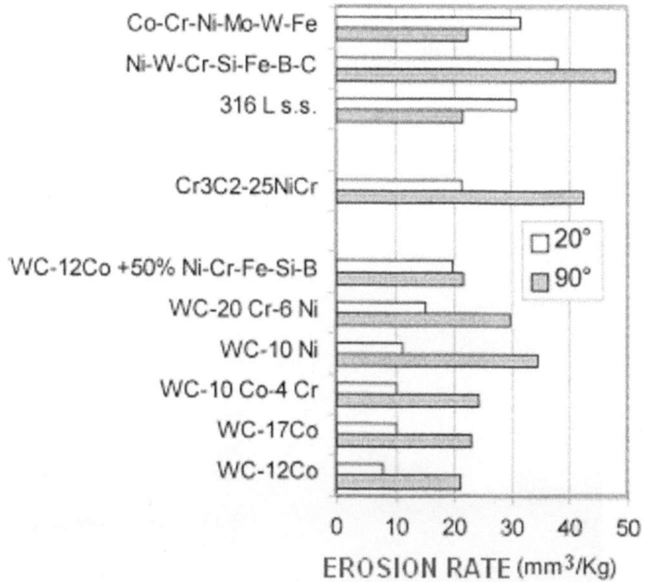

Fig. 27 Erosion rate of different coatings as function of impact angle (50μm particles of Al₂O₃, velocity of 84 m/s, 25 min. test, 50g of erodent, at room temperature) (HAWTHORNE *et al.*, 1999).

Vicenzi (2007) noted that the incorporation of Cr_3C_2 hard phases into the NiCr matrix makes the coating more characteristically brittle for erosion wear, with higher erosion rates at erosion attack angles near 90°, unlike the observed by Hawthorne *et al.* (1999). Vicenzi (2007) also observed that regardless of the amount of the hard phase added, the application process of the coating, or the porosity of the coating, the increase of Cr_3C_2 makes the coating characteristically brittle for erosion. The evolution of microstructure with the increase of Cr_3C_2, after erosion, shown in Figure 30, confirms this statement.

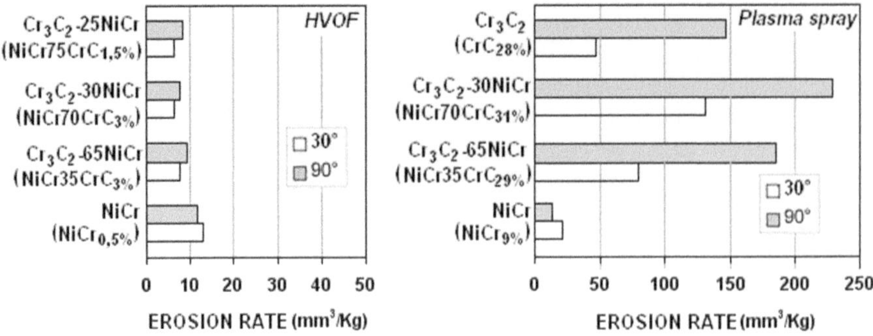

Fig. 28 Erosion rates for different coatings of the NiCr–Cr$_3$C$_2$ system as function of impact angle and amount of ceramic phase (180μm particles of Al$_2$O$_3$, velocity of 50 m/s at room temperature - 10 min test for the plasma spray group and 30 min for HVOF group) (VICENZI, 2007).

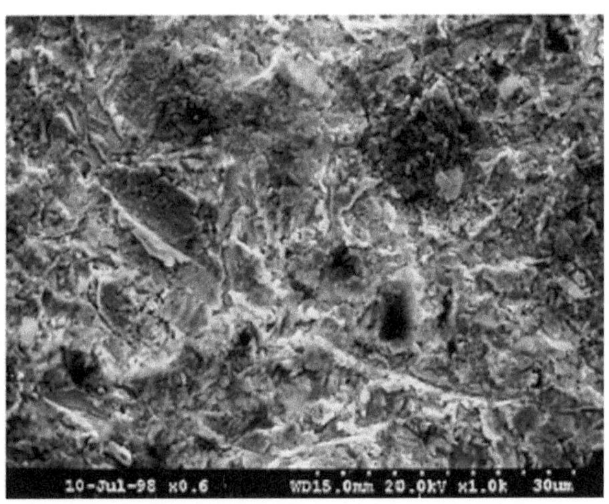

Fig. 29 Typical microstructure of a WC-12Co coating after erosion at 90° impact angle (50μm particles of Al$_2$O$_3$, velocity of 84 m/s, 25 min. test, 50g of erodent, at room temperature) (HAWTHORNE *et al.*, 1999).

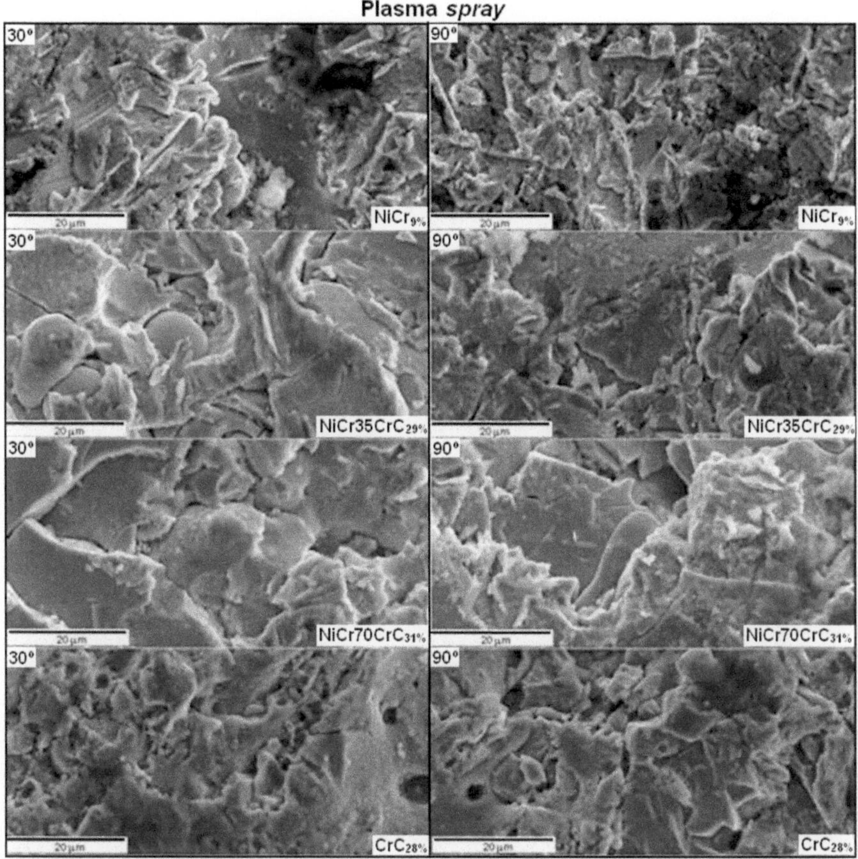

Fig. 30 Typical microstructure of NiCr-Cr$_3$C$_2$ system coatings deposited by plasma spray and HVOF after erosion at impact angles of 30 and 90° (180µm particles of Al$_2$O$_3$, velocity of 50 m/s at room temperature - 10 min test for plasma spray group and 30 min for HVOF group) (VICENZI, 2007).

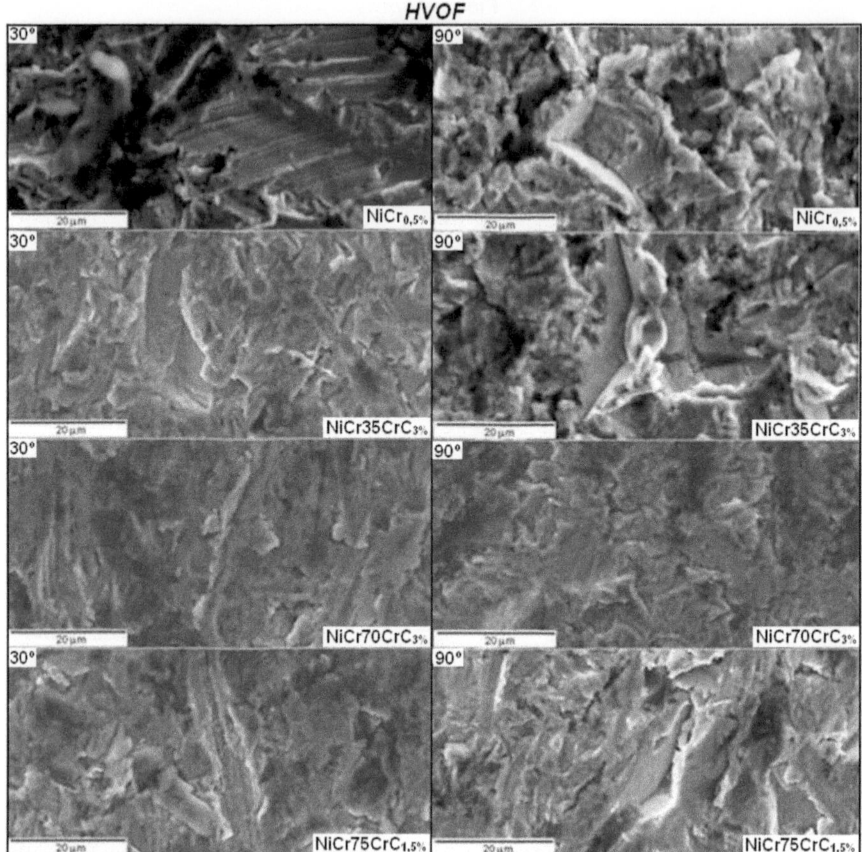

Fig. 30 (*continued*)

5.3 *Amount of Hard Phases (Carbides) Added to the Coatings*

The influence of the amount of hard phases in coatings was also investigated by Kulu and Pihl (2002). These authors observed that in the case of erosion at low- and medium impact angles, the wear rate decreases as the hardness increases, and micro-cutting mechanisms are predominant. The preferential microstructure presents a higher amount of carbide phase (Figure 31a), which must be greater than 50%. In the case of near-normal impacts, a microstructure with higher amounts of ductile matrix and amount of hard phase less than 50% are favorable (Figure 31b). For applications, a combination of impact angles and an excellent microstructure is necessary, for example, a WC-Co coating consists of a ductile matrix with small carbide particles (in this case WC) and agglomerates of WC-Co. This microstructure is called double-cemented (Figure 32a). Another means to obtain this type of microstructure is by aspersion with posterior melting of the coating, as in self-fluxing alloys. An example of this type of alloy with cermet particles is shown in Figure 32b, in this case NiCrSiB with WC-Co.

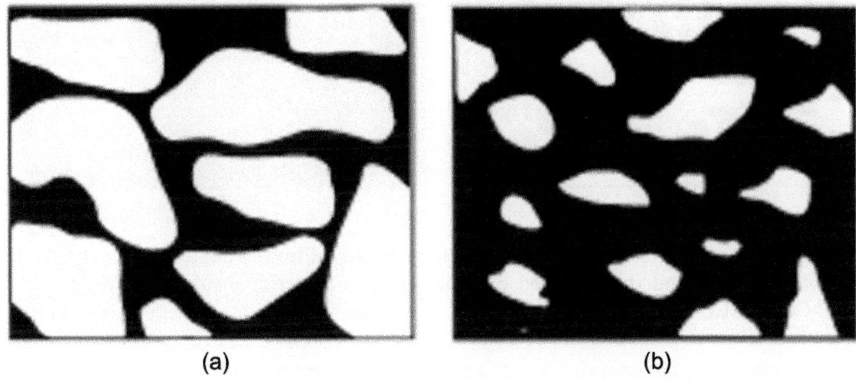

(a)   (b)

Fig. 31 Recommended microstructures to obtain erosive wear strength (a) oblique impact angles and (b) normal impact angle (KULU and PIHL, 2001).
Black= metallic phase; White = hard phase.

Levy (1988) describes in his work that a minimum of 80% of hard phase, such as WC or Cr_3C_2, is necessary to obtain the maximum erosion strength. Lewis *et al.* (1989) report that the higher the Cr_3C_2 content in the powder before application, the higher is the erosion strength of the coating that is obtained. In support of Levy, Walsh and Tabakoff (1990) have shown that a coating made from a powder containing 80% Cr_3C_2 is more erosion resistant than one with 65% Cr_3C_2. In addition, Lewis *et al.* (1989) have shown that the erosion resistance of cermet coatings increases with an increase in chromium carbide in the pre-sprayed powder. However, several studies involving tungsten carbide have shown contrary results. Tu *et al.* (1991) *apud* Stein *et al.* (1999) have shown that an optimum amount of tungsten carbide for erosion resistance, exists at 35% WC in the pre-sprayed powder. Wang *et al.* (1990) *apud* Stein *et al.* (1999) found that as the percentage of tungsten carbide in the coating increased from 7% to 16%, the erosion rate increased. Although a variety of erosive particles were used in the previously mentioned studies (Al_2O_3, SiC, and SiO_2) no trend could be found to explain the differing results reported by the various researchers. As comment in this work many authors present satisfactory results of wear strength of composite coatings, but there is still discussion about a better hard phase/matrix ratio to promote better erosion and erosion–oxidation strength.

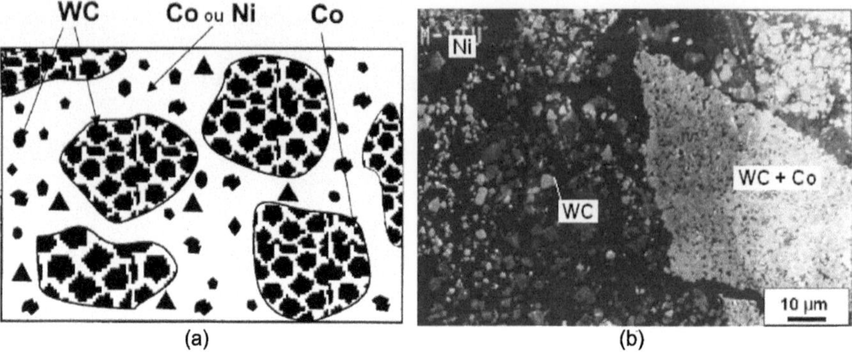

Fig. 32 Ideal microstructure of cermet coatings for varied impact angles, (a) recommended *double cemented* and (b) produced by aspersion and posterior melting (NiCrSiB + 25% WC-Co) (KULU and PIHL, 2001).

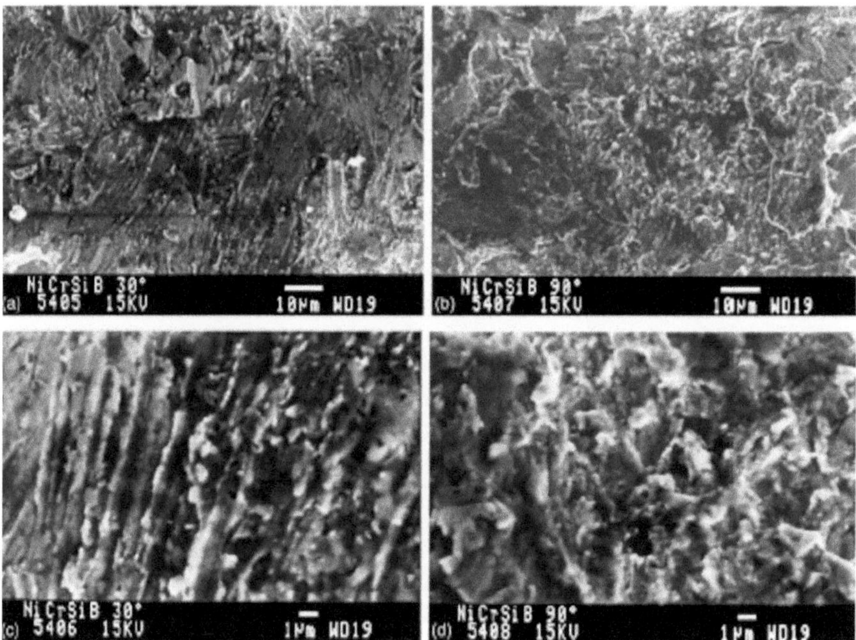

Fig. 33 Surface of the NiCrSiB coating sprayed by *HVOF* eroded at impact of (a) and (c) 30° and (b) and (d) 90° (KULU *et al.*, 2005).

According Stein *et al.* (1999) researchers agree that carbide-based coatings (cermets) provide excellent erosion protection, but disagree on the optimum amount of carbide for maximum erosion resistance. Past studies have used a limited range of pre-sprayed powder carbide contents and failed to fully characterize the final as-sprayed coatings. Thus, Stein *et al.* (1999) evaluated the erosion strength of coatings based on NiCr and FeCrAlY, with content of Cr_3C_2 ranging from 0 to 100%, sprayed by HVOF. These authors observed a low aspersion yield of this material, probably due to flame temperature which causes the recoiling of the particles on the substrate surface, in solid state, not being kept in the coating. The carbide can transform from Cr_3C_2 to Cr_7C_3 and to $Cr_{24}C_6$. Walsh (1992) observed that the carbide oxidation can occur during the application of coating, so the resulting hard phase will be formed by oxides and carbides. In the same work, the maximum erosion strength at impact angle of 90° was achieved with hard phase content of 40%.

Kulu *et al.* (2005) also observed in practice the influence of the amount of carbides, porosity, as well as the impact angle. In this study on three different classes of coatings sprayed by four distinct aspersion techniques (*HVOF*, *D-gun*, *FSF – flame spray fusion*, and *LSF – laser spray fusion*), the authors observed that a comparison can be made for coatings with similar porosities (0.7–3%). So, analyzing the images of the topography of the eroded surface of a NiCrSiB coating sprayed by HVOF and that of WC-17Co also by HVOF at angles ranging from 30 to 90° (Figures 33 and 34, respectively), they concluded that the wear at high impact angles resulted from fracture of carbides or from removal of micro-particles due to a low-cycle fatigue process.

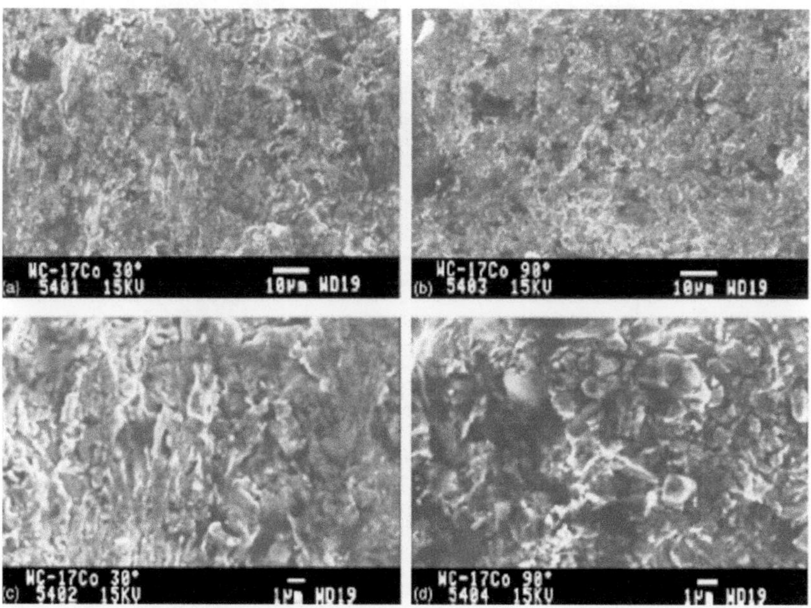

Fig. 34 Surface of the WC-17Co coating sprayed by *HVOF* eroded at impact of (a) and (c) 30° and (b) and (d) 90° (KULU *et al.*, 2005).

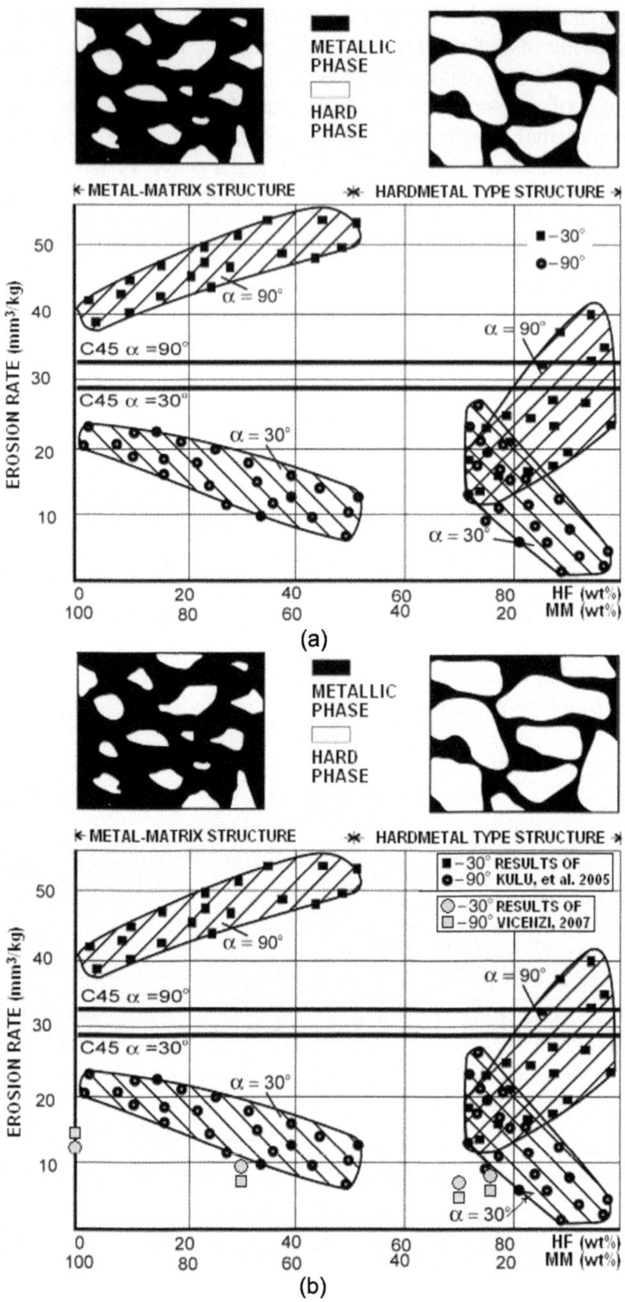

Fig. 35 Erosion rate of solid particles at room temperature as function of the hard phase (HP)/metallic matrix (MM) contents (a) results of KULU *et al.*(2005) (b) results of KULU *et al.*(2005) added to these results the results of Vicenzi (2007).

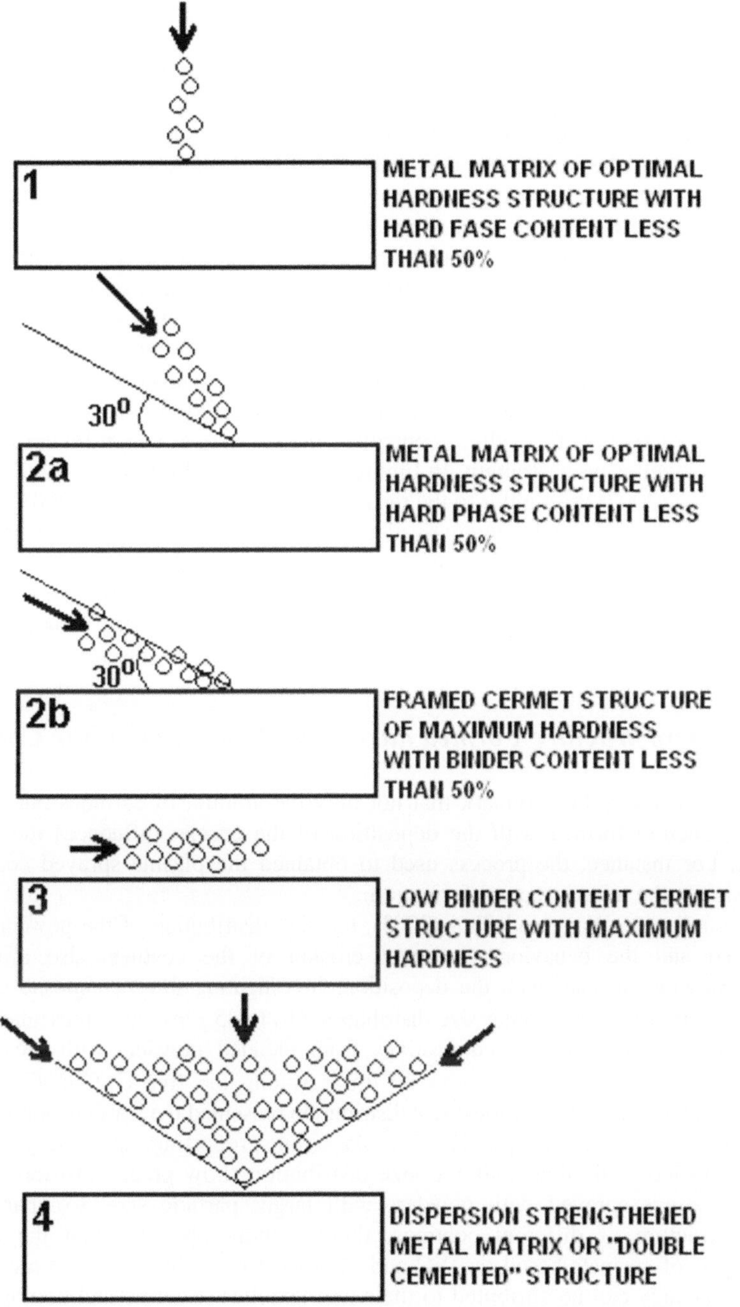

Fig. 36 Optimized structure for resisting erosive wear at different attack angle conditions (1) near normal (90°), (2a) and (2b) oblique, (3) tangent erosion and (4) combination of angles (KULU and PIHL, 2001).

Kulu *et al.* (2005) observed that for low impact angles, the micro-cutting processes are dominant in the metallic matrix of the coating. In the case of coatings with low content of binding (metallic phase), fracture and low-cycle fatigue processes are predominant. On the basis the studied coatings, using WC-Co, NiCrSiB, and NiCrSiB+WC-Co, Kulu *et al.* (2005) suggest a diagram (Figure 35) presenting the influence of the amount of hard phase of the coating on the wear rate of thermal sprayed coatings. Vicenzi (2007) presented similar results as Kulu *et al.* (2005) when analyzing NiCr-Cr_3C_2 system coatings deposited by *HVOF* technique, and the results are shown in Figure 35b. Vicenzi (2007) investigated NiCr-Cr_3C_2 system coatings deposited by plasma spray, but due to its high porosity, could not be compared to the work of Kulu *et al.* (2005).

Figure 35a makes clear that in low impact angles, the wear resistance increases with the hardness increase and the micro-cutting mechanism is dominant. So, a structure with high amount of hard phase is preferred. In the case of normal impact, structures with medium content of hard phase, although higher hardness presents a better wear strength. In Figure 35b, it is noted that the results obtained by Vicenzi (2007) are similar to the results presented by Kulu *et al.* (2005).

Kulu and Pihl (2002) summarized the conclusions drawn by them about the relationship between the impact angle of the erodent particles and the microstructure of the coating with better behavior against the erosion, as a function of the hardness of the coating and the amount of hard phase, which is shown in Figure 36.

5.4 *Microstructure Formed during the Deposition of the Coating*

Wang and Shui (2003) remark that not only the amount of carbides but also the microstructure formed with the deposition of the cermet influences the erosive wear. For instance, the process used to obtained the plasma sprayed coating is decisive in the microstructure formation.

According to Wang and Shui (2003), the size distribution of the powders to be sprayed and the behavior of cermet erosion of the coatings also affect the microstructure formed with the deposition. Investigating three composite coatings based on powder with similar size distribution (16 to 45 μm), these presented higher erosion strength than sprayed coatings from blended powders with greater size distribution (5 to 45 μm or 5 to 80 μm). Among the five Cr_3C_2-NiCr blended powders, the one with the widest size distribution presented the least erosion strength.

Kulu *et al.* (2005) also published similar results, reporting that coatings sprayed with powders with almost similar size distribution show greater erosion strength than coatings sprayed with powders with larger particle size. Comparing the microstructure of different coatings, these authors observed that the erosion strengths of cermets based on WC and of some Cr_3C_2-NiCr coatings are higher. This tendency can be attributed to their fine lamellar structure and homogeneous distribution (Figure 37a) of the hard phases as shown in Figure 37c, as well as the high micro-hardness of the coatings. In the microstructure of other blended powder coatings and coatings sprayed from the Cr_3C_2 cermet, large particles of

NiCr metal distributed randomly and coarse sized lamellas, which are related to lower erosion strength, than the coated and sprayed composite, are present (Figure 37b and 37d) (WANG and SHUI, 2003).

In the investigation of eroded and coated surfaces as well as their cross-sections, the possible mechanisms of erosive wear can be demonstrated. Figure 38 presents the micrographs (SEM) of samples coated with blended WC–17CoCr and Cr_3C_2-NiCr cermet, in which cracking and decrusted morphology are observed, indicating the occurrence of brittle erosion mechanisms (WANG and SHUI, 2003). Levy and Wang (1988) suggest that when the damage by erosion occurs by cracking and decrusting mechanisms and when there is loss of pieces, the size of the extracted material is determined by the size of the lamellas of the coating.

It is suggested that the cracking occurs at first on the lamella boundary during the impact of the particles. With the continuous impact of particles, radial and lateral cracks are developed, leading to fracture and loss of material by decrusting (similar to brittle-material erosion mechanism). Finally, very small voids and pits are formed. It is clear from Figure 38a and 38b that the WC-17CoCr coating presents a fine structure, with small-sized lamellas, which restricted the size of the pieces removed by the impact of particles, contributing to the low damage caused by erosion and a smooth surface morphology. Alternatively, it is observed that the damage generated on the surface of blended Cr_3C_2-NiCr (Figure 38c) was more aggressive, causing a greater loss of surface material, but the mechanism was similar to that observed for WC-17CoCr.

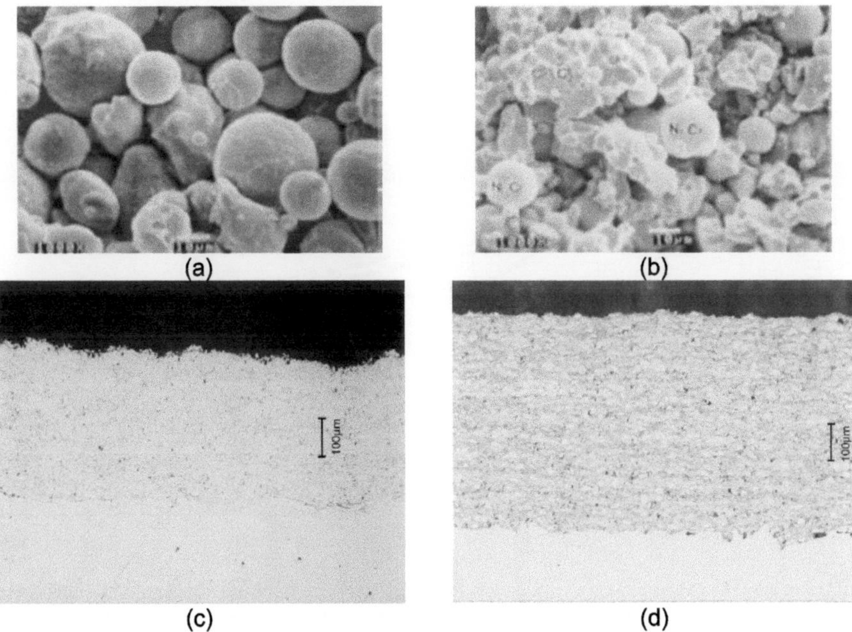

Fig. 37 SEM micrograph of powders before thermal aspersion (a) Cr_3C_2-NiCr – powder and (b) Cr_3C_2-NiCr – blended. Cross-section of coatings (c) Cr_3C_2-NiCr – powder and (d) Cr_3C_2-NiCr – blended (WANG and SHUI,2003).

The effect of carbide grain size in the matrix should also be evaluated, according to Shetty *et al.* (1987). A comparison of the erosion rate of WC-Co cermets with three different carbide sizes, namely, 1.0 μm, 1.4–2.0 μm, and 2.0–3.7 μm, was made. The authors observed that the erosion rate weakly decreases as the grain size of the carbide increases. However, the range of carbide grain size and the variation of erosion rate with the carbide size studied are both small, and, therefore the exact relationship of the carbide grain size with the erosion cannot be established.

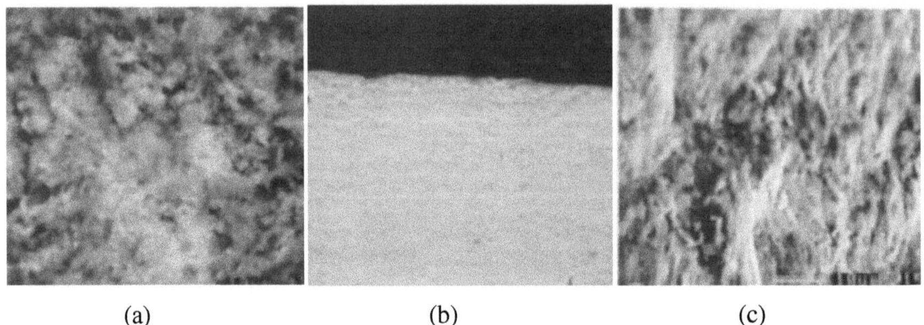

(a)   (b)   (c)

Fig. 38 SEM micrography of (a) surface, (b) cross section of WC-17CoCr coating and (c) surface of Cr_3C_2-NiCr – blended coating eroded by coal ash at angles from 30° to 300° of temperature (WANG and SHUI,2003).

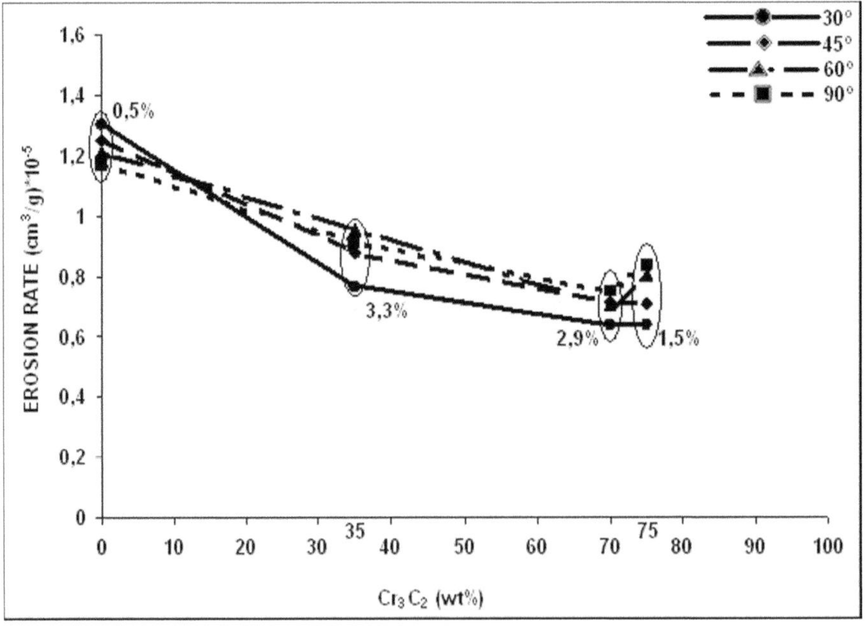

Fig. 39 Erosion rate variation, volume loss by erodent mass impacted, as fuction of the incident angle of erodent, amount of ceramic phase (Cr_3C_2) added to the NiCr matrix and porosity of the coatings (VICENZI, 2007).

Vicenzi (2007) also evaluated the effect of different morphologies of powders used in thermal aspersion, aiming to compare the erosive wear after the coatings were obtained, as well as the amount of hard phase (carbide) added. Figure 39 shows the results obtained by the author.

Vicenzi (2007) emphasizes in Figure 39 that the erosion rate reaches the lowest values when the amount of Cr_3C_2 added to NiCr is increased by 70% at all attack angles analyzed (30, 45, 60 and 90°), unlike the observation made by Kulu *et al.* (2005), as commented before. According to those authors at low impact angles, the wear strength increases as the hardness increases, so, a structure with high content of hard phase is preferred. In the case of normal impact, a structure with medium content of hard phase, although higher hardness, shows a better strength wear. However, these authors did not consider the morphology of the powders which generate the coatings, and, in this sense, Vicenzi (2007) emphasizes that an increase of Cr_3C_2 by 75% resulted in a higher erosion rate (at attack angles of 60° and 90°) and a higher dispersion of results as function of the attack angle, which can be linked to the type of the powder used in HVOF aspersion, disagreeing with the expected results given by Kulu *et al.* (2005). This was possibly attributed to the fact that the powders used for coating with increase of 35 and 70% of Cr_3C_2 ($NiCr35CrC_{3\%}$ and $NiCr70CrC_{3\%}$) are activated, whereas for the coating with 75% of Cr_3C_2 ($NiCr75CrC_{1,5\%}$), the powder is agglomerated and sintered. In activation, according to the powder manufacturer (PRAXAIR), the chromium carbide particles are coated individually by the nickel matrix, which is probably done to achieve a greater homogeneity of the material and powder fluidity, favoring the increase in deposit efficiency, as well as for optimizing the coating properties.

Matthews, James and Hyland (2009) evaluated the influence of deposition technique on the erosive wear at high temperature (700°C and 800°C). The techniques used were HVOF and HVOA in a agglomerated/sintered powder. In addition, the coatings were thermally treated at 900°C during 2 days and 30 days. The authors observed that the effect of the deposition technique was most evident at 800°C, due to a variation of carbide dissolution in-flight. The HVOF coating presented higher carbide dissolution leading to more ductile erosion (lower erosion rate at 90°) relative to HVAF coating.

In the HVAF erosion the authors observed that the heat treatment led to a higher erosion rate, due to the formation of a three dimensional carbide network, resulting in severe constraint of the NiCr phase and a more brittle erosion response of the cermet. Moreover, the HVOF coating exhibited a consistent erosion rate with heat treatment, with the erosion rate increasing with the heat treatment time, due to the carbide development (MATTHEWS, JAMES and HYLAND, 2009 *apud* LAUGIER, HYLAND and JAMES, 2003 and MATTHEWS, HYLAND and JAMES, 2004). Fine carbide grains in the extensive supersaturated NiCr regions of the coating were observed, which hardens the NiCr matrix, restricting the extent of ductile deformation under impact (MATTHEWS, JAMES and HYLAND, 2009).

At 700°C the authors did not observe a trend in erosion rate as a function of heat treatment, even whenthe same microstructural changes occurred in the HVAF and HVOF coatings. This is caused by the low degree of matrix phase ductility at this temperature which promotes a brittle erosion response in the cermet. The complex relationship between the degree of matrix ductility and the effect of the developing carbide microstructure accounts for the consistent erosion rate of the heat treated samples with that of the as-sprayed samples for both coating sets (MATTHEWS, JAMES and HYLAND, 2009).

Regarding the microstructure after erosion, Matthews, James and Hyland (2009) observed the same erosion features on the as-sprayed HVAF coating eroded at 700°C and 800°C, only more accentuated and at the higher temperature. They cited: i) localised mass loss – fracture of flakes and platelets displaced from the periphery of the indents and ii) extensive mass loss – "gross chipping" mechanism (occurs via a low cycle fatigue mechanism to produce chips of debris of the same magnitude as the depth of erodent indentation), see Figure 40.

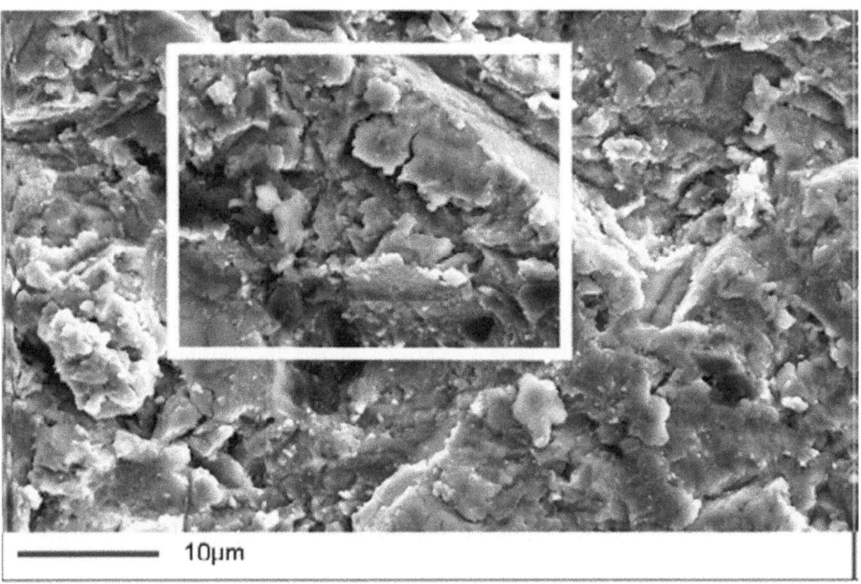

Fig. 40 SEM micrography after erosion of HVAF coating 75 Cr_3C_2-25(Ni(20Cr))as-sprayed showing the gross chipping mass loss mechanism at 800°C.(MATTHEWS, JAMES and HYLAND, 2009).

The erosion observed in as-sprayed HVOF coating was less ductile than in the HVAF coating at both temperatures. Figure 41 shows the less ductile deformation in some single impact indents around the periphery of the steady state erosion crater.

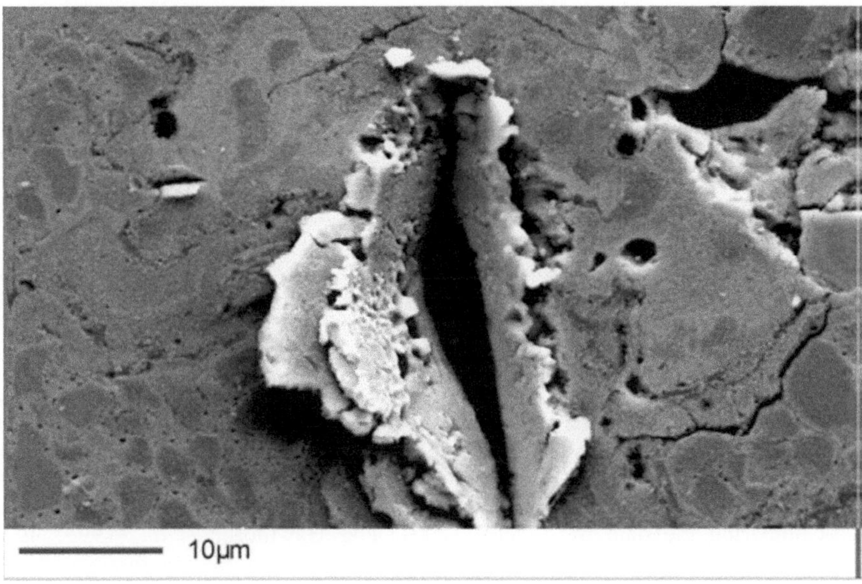

Fig. 41 SEM micrography after erosion of HVOF coating 75 Cr₃C₂-25(Ni(20Cr)) as-sprayed showing the erosion impact deformation at 800°C. (MATTHEWS, JAMES and HYLAND, 2009).

5.5 *Hardness of the Coating*

In the investigation for defining the mechanisms that are responsible for erosion in cermet materials, some authors point out that hardness is a dominant factor (D'Errico *et al.*, 1997). For example, according to Hawthorne *et al.* (1999), for cermet coatings at low attack angles, there is an increase of erosion strength with the increase of the carbide amount and, therefore, higher hardness phase of the coating, probably as function of the mechanism which rules this type of degradation. From Figure 42a, Hawthorne *et al.* (1999) affirm that there is no relation between hardness and erosion strength. Matthews, James and Hyland (2009) agree with these authors, they observed that the coating microhardness measured at ambient temperature is a poor indicator of the relative erosion response of these coatings under aggressive erosion conditions at elevated temperature. However, it can be noticed that for low impact angles, the erosion strength increases with the increase of carbide hardness. Whereas the opposite is the case for metallic coatings, probably due to the different erosion mechanisms which is dominant for each type of these materials.

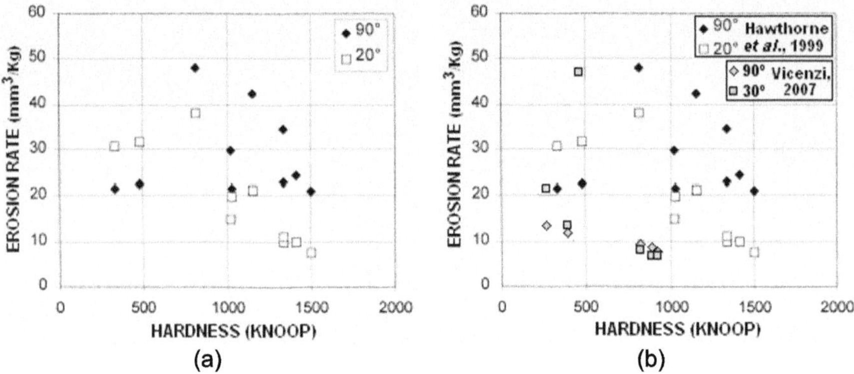

Fig. 42 Relationship between erosion rate and hardness for different coatings tested as function of the impact angle (a) results byHawthorne *et al.*(1999)and (b) comparison of results between Hawthorne *et al.*(1999)and Vicenzi (2007).

Comparing the data by Vicenzi (2007) and Hawthorne *et al.* (1999), it is observed that actually at low attack angles, the erosion strength increases with the increase of the hard phase; however, this was true for low-porosity coatings (Figure 42b). For coatings with porosities higher than 4%, according to Vicenzi (2007), it is not possible to compare the hardness with the erosion strength (Figure 43). It can be observed in Figure 43 in plasma sprayed coatings that the erosion rate depends of the porosity, without to present some behavior with the hardness. This occurs, probably, due to the low cohesion force between lamellas, with regards to the high porosity of the coatings. So, not only the hardness of the coating is a determinant factor to its erosive wear strength, but other factors, such as its microstructure and the mechanism responsible for the wear can be determinant (VICENZI, 2007).

In agreement to the work by Vicenzi (2007), Kulu *et al.* (2005) summarize the influence of the hardness on the wear strength of cermet coatings. These authors observe that at room temperature, the hardness has a great effect on material wear by plastic deformation mechanism, whereas the fracture tenacity is a dominant factor in case of wear involving brittle fracture. However, other properties are also involved in the determination of erosive wear forms of cermet coatings, such as, for example, their porosity.

Kulu *et al.* (2001), Kulu and Veinthal (2000) *apud* Kulu and Pihl (2002) correlated the hardness of coatings sprayed with different processes with erosion rate and the impact angle. These authors observed that for coatings deposited by HVOF at low impact angles (30°), the wear decreases with increase of coating hardness, due to a micro-cutting dominant mechanism, which was in agreement with the results obtained by Hawthorne *et al.* (1999) and Vicenzi (2007). At high impact angles (90°), increasing the coating hardness up to 700–800 HV causes an increase in the wear rate. For more hard coatings (WC-Co thermal sprayed by *HVOF*), an increase in coating hardness causes a decrease in the wear rate, due to the predominance of low-cycle fatigue fracture and direct fracture mechanisms. This dependency of the hardness on the impact angle and erosion rate is presented in Figure 44.

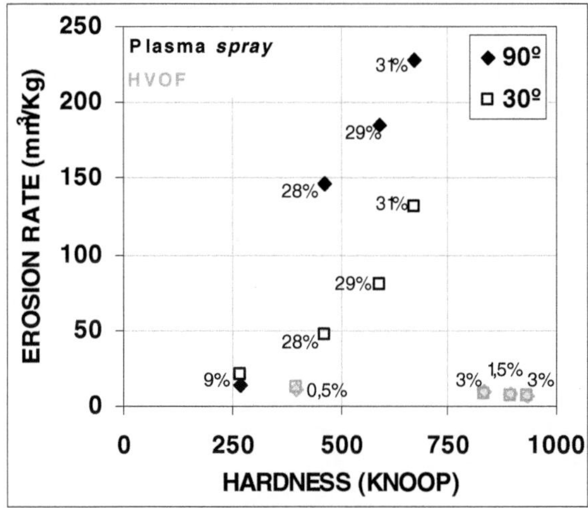

Fig. 43 Relationship between erosion rate and hardness for different coatings from the system NiCr-Cr$_3$C$_2$ depositedby*HVOF* (low porosity) and plasma *spray* (high porosity) (VICENZI, 2007). The percentual value are the porosity of sprayed coatings.

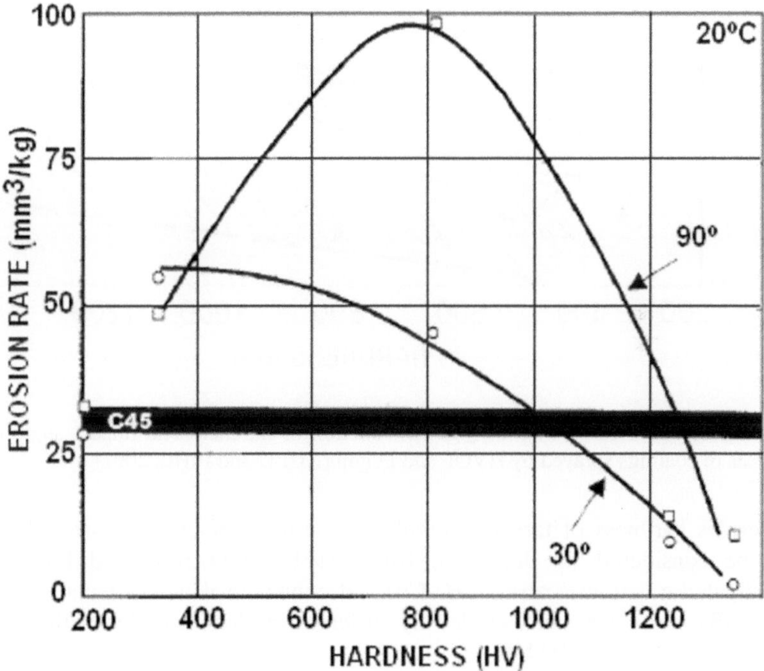

Fig. 44 Dependency of erosion rate as function of hardness and impact angle of the particles (dark area depicts the reference material: 1045 steel) (KULU and PIHL, 2001).

Figure 45 presents the dependency of the relative wear strength of coatings sprayed by HVOF and D-gun on hardness and impact angle. It is noted that with the increase in the hardness of HVOF coatings, the erosive wear strength increases for both low and high impact angles.

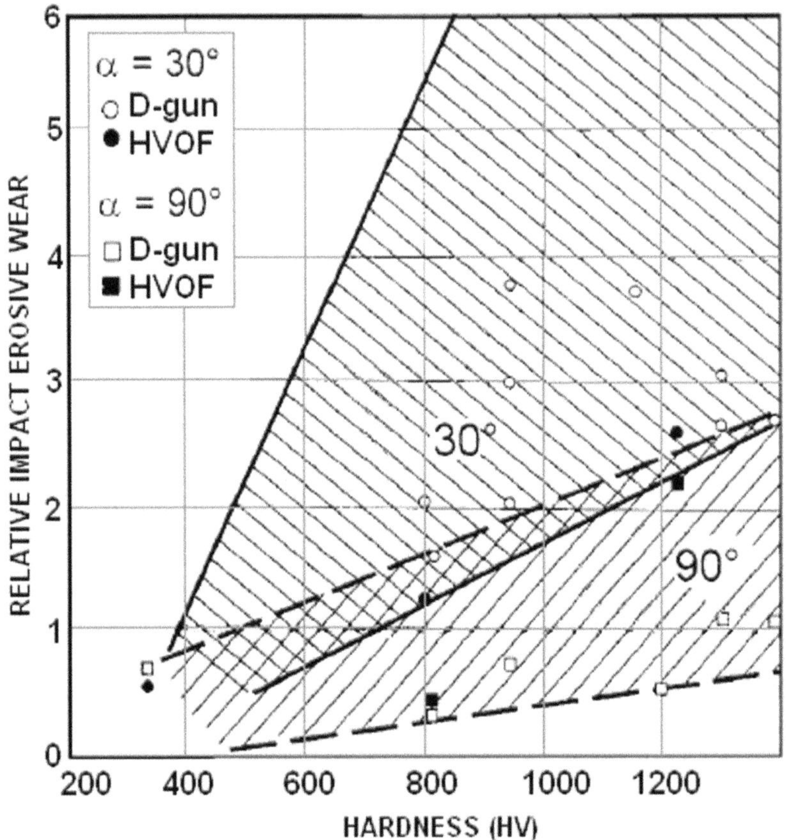

Fig. 45 Relative erosive wear strength as function of hardness and impact angle of the particles of coatings sprayed by HVOF and D-gun (KULU and PIHL, 2001).

Besides hardness of target material, the hardness of erodent material should also be considered. In doing so, Kulu (1989) *apud* Kulu and Pihl (2002) investigated a cermet coating (WC-Co) and compared it with a metallic coating (NiCrSiB) and 1045 steel, eroded by different erodent particles with hardness ranging from 120 to 2000 HV.

The dependency of erosion rate as function of the impact angle and the erodent hardness is shown in Figure 46, in which it is observed that in order to maintain the high wear strength at oblique angles, the hardness of coating should exceed the erodent hardness. For normal impacts, as fracture by fatigue mechanisms prevail, the influence of the erodent hardness is insignificant.

Vicenzi (2007) also emphasizes that the initial hardness of the coating should be considered, as well as the change in this property due to the hardening of the eroded surface. This hardening is caused by the successive strikes of erodent particles over the metallic part of the cermet, resulting in a possible fragilization of the matrix. Thus, the carbide bonding becoming less effective and the wear mechanism becoming more predominant.

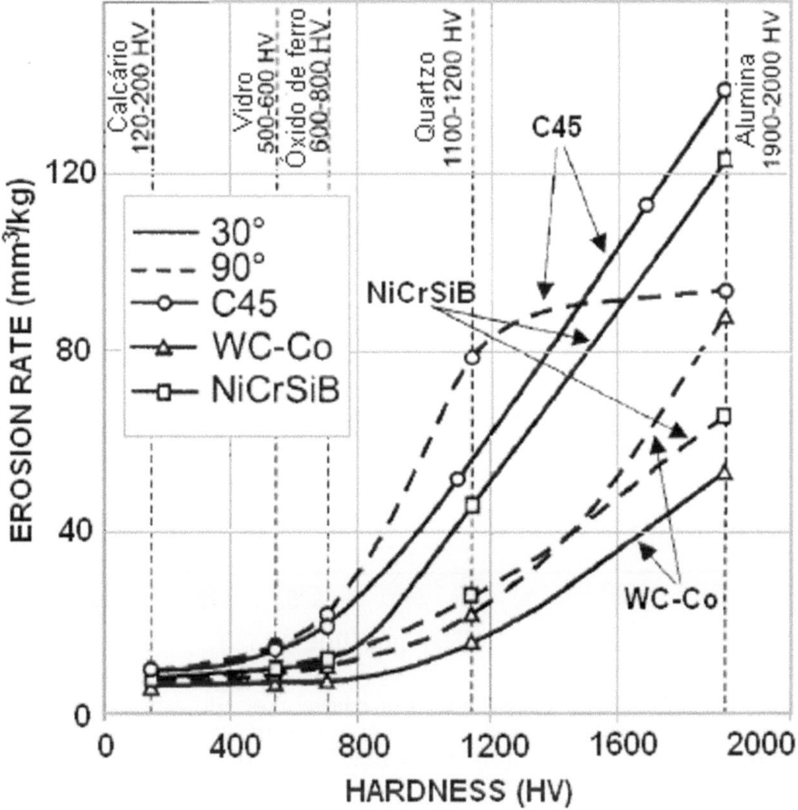

Fig. 46 Variation of the erosion rate with the hardness of erodent particles as function of particle impact angles for coatings sprayed by HVOF (KULU and PIHL, 2001).

5.6 *Temperature in Erosive Wear*

Similar to metals, an increase in temperature can cause changes in the mechanical properties of the coatings, as well as in their erosion mechanism and even a possible oxidation.

Wang and Verstak (1999) observed the relationship between erosion rate and temperature of two different coatings sprayed by HVOF (Cr_3C_2/TiC-NiCrMo and Cr_3C_2-NiCr), as shown in Figure 47. The same tendency is observed for both the investigated coatings: until 300°C, loss of thickness (as criterium for erosion damage) decreases and from 300°C to 750°C, it increases. Tayor *et al.* (1997) *apud* Wang and Verstak (1999) obtained similar results for alumina and zirconia coatings, sprayed by plasma spray.

It is also clear from Figure 47 that the Cr_3C_2/TiC-NiCrMo coating is more sensitive to temperature than the Cr_3C_2-NiCr. Below 600°C, the loss of thickness of the Cr_3C_2/TiC-NiCrMo coating is less than that of Cr_3C_2-NiCr. However, as the temperature increases, the Cr_3C_2/TiC-NiCrMo coating loses more thickness (at a rate higher than the loss by Cr_3C_2-NiCr coating). This possible greater loss of Cr_3C_2/TiC-NiCrMo coating can be related to the TiC oxidation.

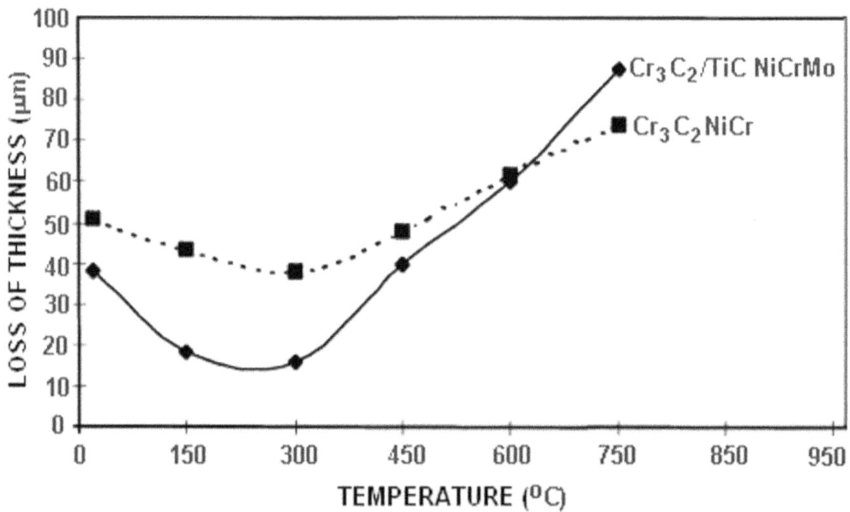

Fig. 47 Loss of thickness due to erosion of mineral coal ash particles as function of temperature for coatings sprayed by HVOF (Cr_3C_2/TiC-25(Ni40CrMo) and Cr_3C_2-25(Ni44CrSi)) at impact angles of 30° (WANG and VERSTAK, 1999).

Figure 48 presents the Cr_3C_2/TiC-NiCrMo coating surface eroded at 30° and 90° by angular particles of mineral coal, at temperature of 300°C and velocity of 60 m/s. The coating eroded at 30° presented brittle behavior, as can be observed from its cracked and decrusted surface (Figure 48a). According to the authors, analyzing the cross-section of the microstructure (Figure 48b), it is possible to observe lateral and radial cracks, as well as the removal of parts of the coating. In the coating eroded at 90°, a fine morphology is observed, with incrustation of erodent particles and presence of small cracks and decrusting (Figure 48c and 48d).

According to Wang and Verstak (1999), the oxidation of coating could be determined by EDS analysis, shown in Figure 49, depicted by a layer of the product of the oxidation over the coating surface (Figure 49a). The TiC coating was eroded at 30° by angular particles of mineral coal ashes, at temperature of 750°C and velocity of 60 m/s. In Figure 49b, comparing the internal with the superficial EDS of the coating, as the Ti content increases, the product of oxidation can be found on the surface, meaning that more Ti was oxidized or corroded at this temperature, according to the same authors.

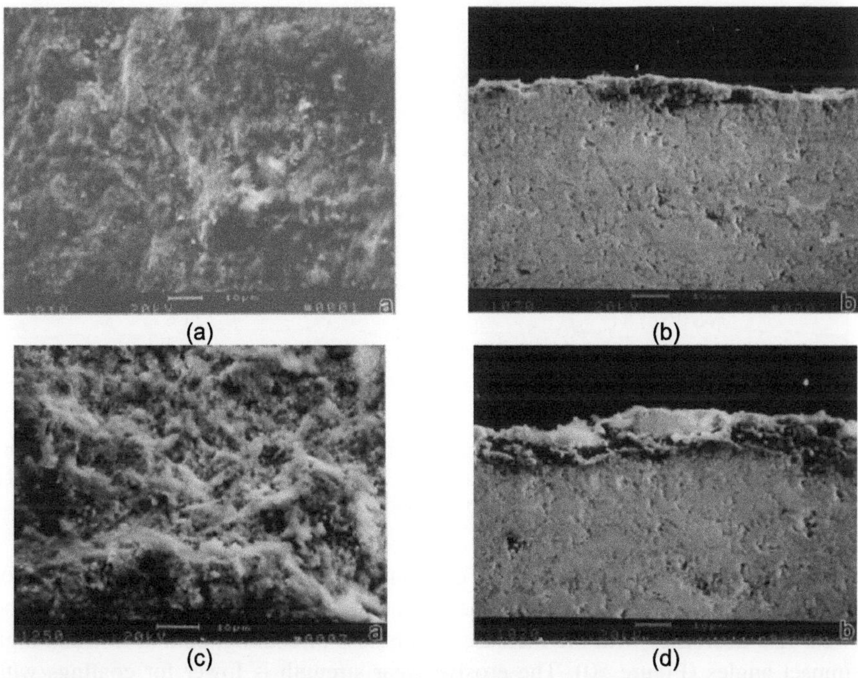

Fig. 48 Microstructure of surface and cross-section of Cr_3C_2/TiC-NiCrMo coating eroded at impact angle of (a) and (b) 30° and (c) and (d) 90°, velocity of 60m/s, and temperature of 300°C (WANG and VERSTAK, 1999).

Besides oxidation at high temperatures, Levy and Wang (1988) propose the possible modification of the energy balance, due to modification of the properties of the target material. These authors observed that the coatings with higher content of metal presented more tolerance to erosive wear at high temperatures (500°C) for velocity of 70 m/s, which is due to the fact that the metallic component dissipates the impact force of the particles by plastic deformation and, then, reduces the erosion rate. Both materials, essentially brittle, Cr_3C_2 and WCNiCrB, presented higher erosion rates for lower velocities due to large decrusting of the materials that are detached from their surfaces by large-sized and angular particles. The texture of the eroded surface was proportional to erosion rate, and the coarser the surface, the higher is the erosion rate.

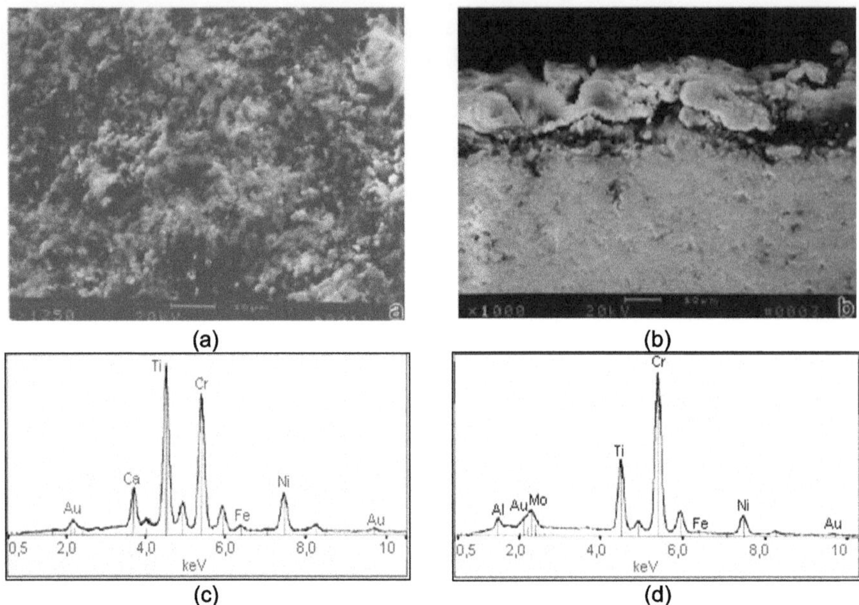

Fig. 49 Microstructure of (a) surface and (b) cross-section of Cr_3C_2/TiC-NiCrMo coating eroded at impact angle of 30° at temperature of 750°C and EDS analysis (c) of the surface (oxidized) and (d) of the coating core (no oxidation) (WANG and VERSTAK, 1999).

Kulu *et al.* (2005) compared the erosion rate at high temperatures (700°C) of coatings obtained by HVOF with the hardness. These authors confirmed that the wear rate decreases with the increase in hardness of coating, both at low and high impact angles (Figure 50). The erosive wear strength is lower for coatings with lower hardness. The wear mechanism of the coatings at high temperature differs from that at room temperature.

At 700°C, the wear mechanism at impact angles of 30° and near 90° are similar. From Figure 51, the erosion caused by ploughing of the erodent particles on the target material can be observed.

In another research, Kulu and Veinthal (2000), Kulu *et al.* (1999), *apud* Kulu and Pihl (2002) also observed that at high temperatures (600 a 800°C), the erosive wear mechanism is similar at impact angles of 30° and 90°, as presented in Figure 52. Under both the wear conditions, the authors noted that the erosion caused by ploughing of the erodent material results in material removal.

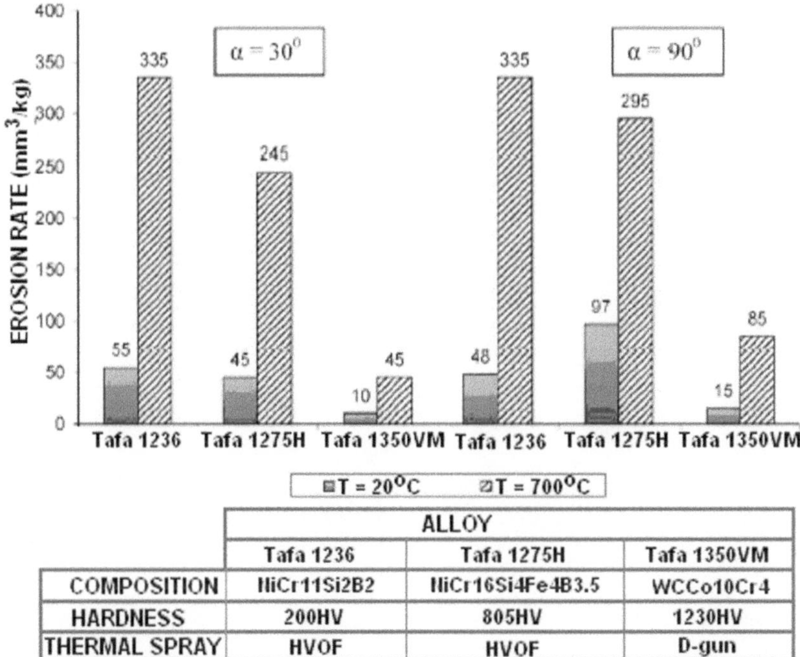

	ALLOY		
	Tafa 1236	**Tafa 1275H**	**Tafa 1350VM**
COMPOSITION	NiCr11Si2B2	NiCr16Si4Fe4B3.5	WCCo10Cr4
HARDNESS	200HV	805HV	1230HV
THERMAL SPRAY	HVOF	HVOF	D-gun

Fig. 50 Erosion rate of coatings as function of the impact angle and temperature (KULU *et al.*, 2005).

According to Vicenzi (2007), the increase in the temperature during the erosion causes the oxidation of the coatings, as well as the increase of the incrustation, which is also observed in coatings of the NiCr-Cr$_3$C$_2$ system at room temperature. The increase of incrustation of erodent particles, according to the author occurs due to a possible increase in the plasticity of the coatings, in agreement with the results of Hidalgo (2001), and their porosity. Furthermore, the author states that the increase of carbide phase decreases the incrustation of the erodent.

Vicenzi (2007) related that the oxidation of the coatings of the NiCr-Cr$_3$C$_2$ system can generate chromium and nickel oxides. So, through image mapping, the author observed that the chromium oxide formation overcomes the nickel oxide formation. Moreover, the oxide formation ratio is directly influenced by the porosity, surface area, and by the amount of Cr$_3$C$_2$ added to the coatings. For the study of the oxidation of coatings of the NiCr-Cr$_3$C$_2$ system and the influence of the factors mentioned above, Figure 53 shows a comparison of the mass gain of coatings obtained by plasma spray and that obtained by HVOF, where it can be noted that a greater mass gain was observed in the coatings obtained by plasma spray. Besides that, cermets are in an intermediary position with respect to mass gain due to the oxidation between phases of only NiCr or only Cr$_3$C$_2$, which is in agreement with the suggestion of the author. Concerning the amount of Cr$_3$C$_2$ phase added to the NiCr matrix, it is noted that the increase of this phase leads to an increase in oxidation of the coatings.

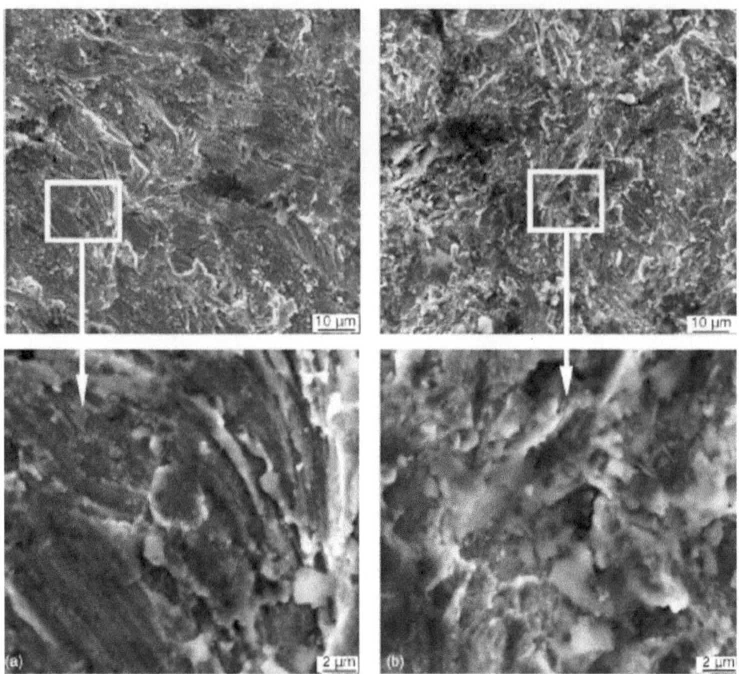

Fig. 51 NiCrSiB coating eroded at temperature of 700°C and impact angle of (a) 30° and (b) 90° (KULU*et al.*, 2005).

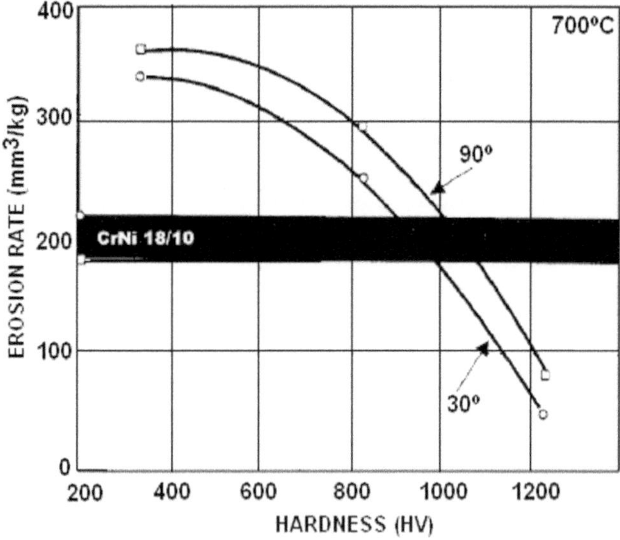

Fig. 52 Variation of erosion rate as function of hardness and impact angle of the particles at temperature of 700°C (dark area depicts the reference material: 18/10 stainless steel) (KULU andPIHL, 2001).

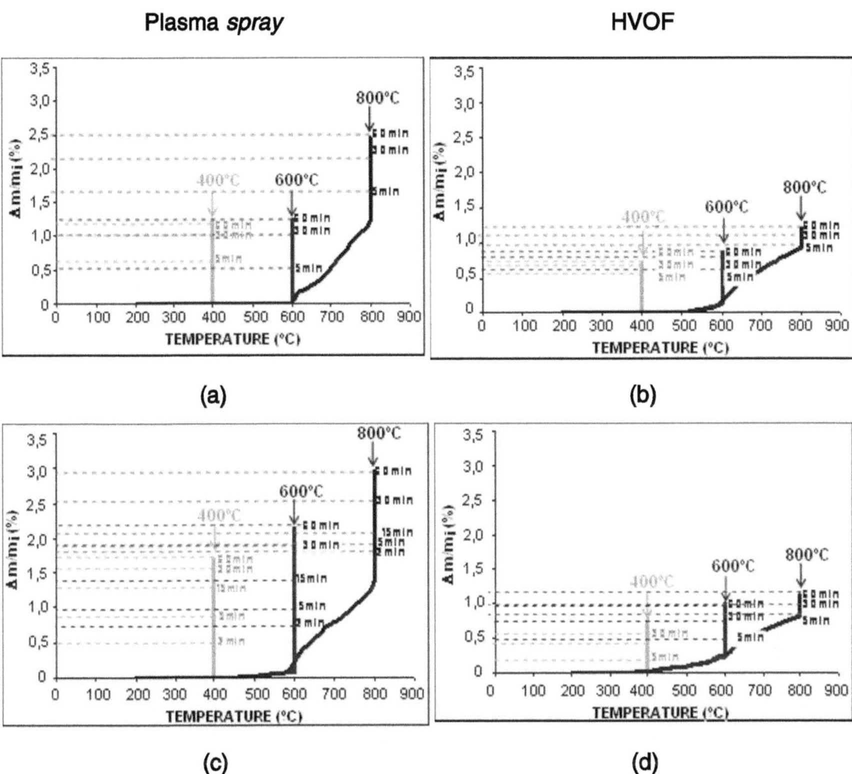

(a)   (b)   (c)   (d)

Fig. 53 Mass gain (wt%) as function of time and temperature of oxidation of the coatings obtained by plasma spray (a) $NiCr35CCr_{29\%}$, and (c) $NiCr70CCr_{31\%}$ and obtained by HVOF (b)$NiCr35CCr_{3\%}$, and (d) $NiCr70CCr_{3\%}$ (VICENZI, 2007).

Vicenzi (2007) observed the behavior of coatings of the $NiCr$-Cr_3C_2 system with respect to erosion rate intensity as function of temperature and attack angle (Figure 54).

For coatings with porosity higher than 4% (obtained by plasma spray), the author observed that the increase in temperature from 25°C to 400°C leads to a decrease in the erosive wear rate for all the analyzed coatings (VICENZI, 2007). This drop is greater for the coating with higher amount of Cr_3C_2 phase. Similar results were obtained by Wang and Verstak (1999). With regards to the attack angles, it is observed that a greater drop occurred at angle of 90° when compared with that at 30°. So, it is supposed that, increasing the temperature leads to an increase in the ductility of coatings with metallic phase, leading to an increase in the erosion rate at lower angles, as described by Hutchings (1977).

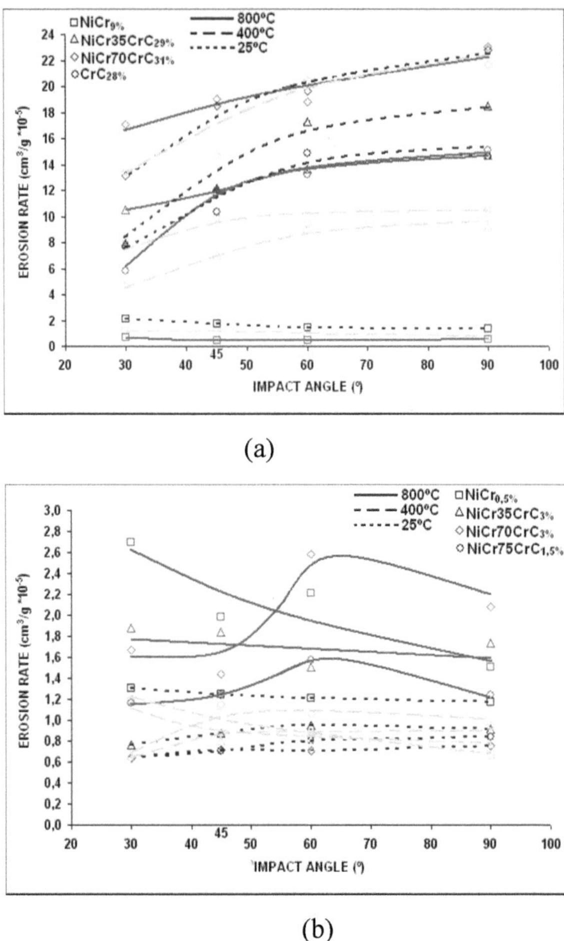

(a)

(b)

Fig. 54 Variation of the erosion rate in volume as function of the erodent attack angle, normalized by temperature and type of coatings deposited by (a) plasma *spray*and (b) *HVOF*(VICENZI, 2007).

Likewise, for coatings with porosity less than 4% (obtained by HVOF), Vicenzi (2007) observed that the increase of Cr_3C_2 decreases the ductile character of the erosion, being more evident at temperature of 800°C. For coatings with 70 and 75% of Cr_3C_2, it is noted that at attack angles of 45°, 60°, and 90°, there is an increase of the erosive wear rate, leading to similar curves as those presented due to the wear of ceramic materials, at 400°C, and more evidently at 800°C. This fact is in agreement with Hawthorne (1999) who observed the erosion of cermets' dependency on the attack angle and on the amount of carbides added to the matrix. Generally, the increase in carbide concentration and attack angle lead to an increase in the erosion strength at room temperature. At high temperatures, Kulu *et al.* (2005) verified that the wear rate decreases with the increase of coating hardness, at both low and high attack angles.

Regarding the wear mechanisms observed at high temperatures, Vicenzi (2007) analyzed the microstructure after erosion at 800°C on coatings obtained by plasma spray and by HVOF (Figure 55). Figure 55 (plasma spray) shows that the temperature increased the plasticity only for the pure metallic coating (NiCr$_{9\%}$). For the other coatings, the increase of Cr$_3$C$_2$ phase makes them gradually more brittle at attack angle of 90°. At 30°, cutting mechanisms are observed for coatings with 30% of Cr$_3$C$_2$. The increase of ceramic phase to 70% resulted in a typically brittle wear of the coating, despite the increase in temperature, which is in agreement with the findings by Shipway and Hutchings (1996). From Figure 55 – (HVOF), the author noted the presence of possible oxidation of the coatings and signs of ductile erosion mechanisms up to 35% of Cr$_3$C$_2$. The increase of the ceramic phase to 70% and 75% does not modify the morphology of the coatings eroded at 30° (characteristically ductile), and at 90°, the presence of some sharp corners is noted.

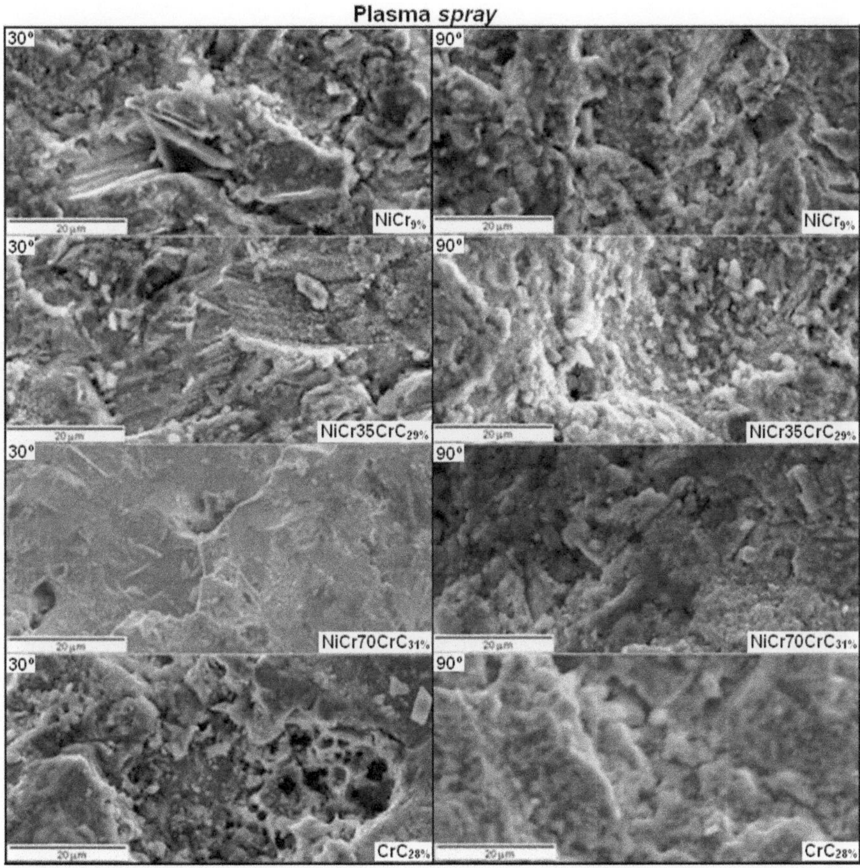

Fig. 55 Micrographs of surface of coatings obtained by plasma *spray*and HVOF after erosion at temperature of 800°C at attack angles of 30° and 90° (VICENZI, 2007).

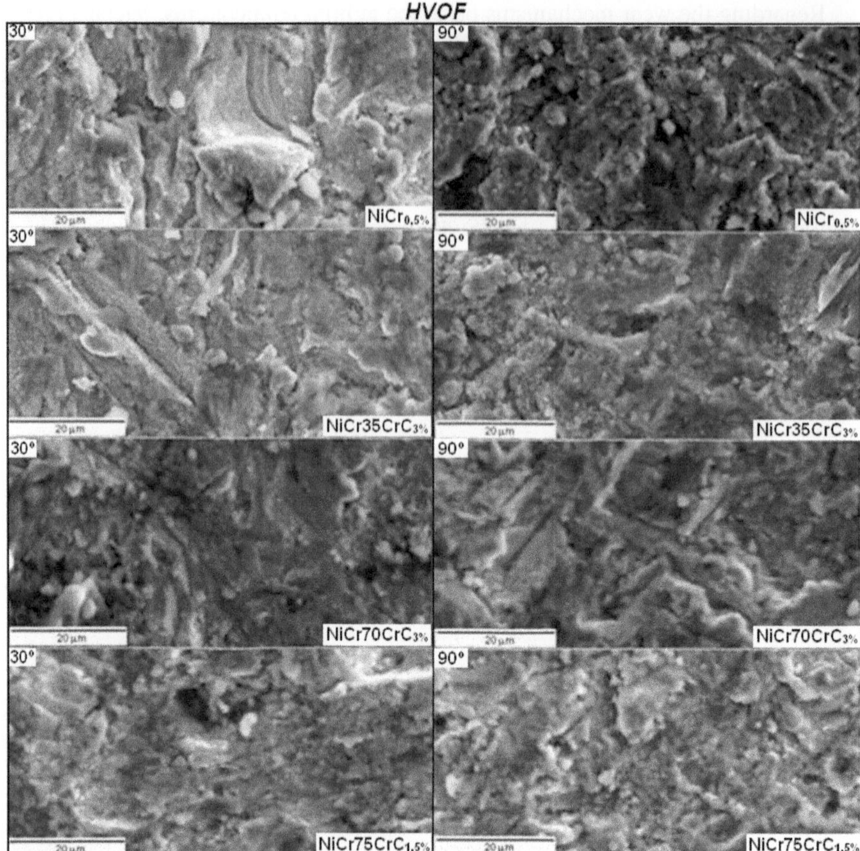

Fig. 55 (*continued*)

6 Summary of the Erosion Mechanisms

Table 6 summarizes the conclusions drawn for erosive wear mechanisms by the researchers mentioned in the review. Table 7 summarizes the results obtained by Vicenzi (2007) in the study on erosive wear at low and high temperatures of NiCr-Cr_3C_2 system coatings.

Table 6 Summary of erosive wear mechanisms mentioned in bibliography review.

Material	Angle	Temperature	
		Low (25°C)	High

Metal	Low	a) Hutchings (1979), Hutchings (1981a), Hutchings (1981b), Cousens and Hutchings (1983), Hutchings (1989) and Finnie (1995) metallic materials by:
		  i) *cutting* mechanism (types I and II)
		  ii) *ploughing*
		  iii) formation and detachment by *platelets*
		
		b) Finnie (1995) – metal (Al, Mg, Au, tool steel) eroded with rounded SiC at angles less than 20º: decrusting.
		
		c) Finnie (1995) – metal (Al, Mg, Au, tool steel) eroded with rounded SiC at angles near 30º: *ploughing.*
			a) Levy *et al.* (1989) – chromium alloy steel eroded with Al_2O_3 rounded at 850°C: erosion by cutting.
			
			b) Levy *et al.* (1989) – erosion of chromium alloy steel with SiC at 850°C: erosion/oxidation mechanisms.

	High	a) Hutchings (1979), Hutchings (1981a), Hutchings (1981b), Cousens and Hutchings (1983), Hutchings (1989) and Finnie (1995) erosion of metallic materials by:
		  i) detachment by *platelet*
		
		b) Finnie (1995) – metal (Al, Mg, Au, tool steel) eroded with rounded SiC at angles near normal: proposed mechanisms:
		  i) brittle behavior of the metal due to the hardening of the surface;
		  ii) particle fragmentation;
		  iii) low cycle fatigue.
		
		c) Cousens and Hutchings (1983) – Al alloy eroded with rounded particles – modified by extrusion of plates by *platelet* mechanism.
			a) Finnie (1995) – metal (Al, Mg, Au, tool steel) eroded with rounded SiC at angles near normal: proposed mechanisms:
			  i) temperature effects due to high deformation rates;
			  ii) wear by delamination and extrusion.
			
			b) Levy *et al.* (1989) – chromium alloy steel eroded with rounded Al_2O_3 at 850°C: erosion/oxidation mechanisms.
			
			c) Levy *et al.* (1989) – erosion of chromium alloy steel with SiC at 850°C: erosion/oxidation mechanisms.

Table 6 (*continued*)

Material	Angle	Temperature	
		Low (25ºC)	**High**
Ceramic	Low	a) Soderberg *et al.*(1981) – aluminas (99.7, 99, and 94%) eroded at 66 m/s: intergranular and transgranular fracture mechanisms. b) Zhou and Bahadur (1995) – alumina+glass eroded with SiC: erosion by formation of decrusted pieces throughout the grain boundary.	a) Marques (2006) – aluminas with and without vitreous phase eroded by electrofused alumina: pit formation with rounded edges and ductile fracture, formation of some cracks. b) Zhou and Bahadur (1995) – alumina+glass eroded by SiC at 650ºC and angle of 10º: groove formation (due to extrusion, *ploughing,* and cutting actions) and few brittle cracks (fragmentation of the material).

	High	a) Madruga, Silveira and Bergmann (1994) – suggested mechanisms at normal impact:
		i) fatigue of intergranular phase,
		ii) microcracks in grain boundary,
		iii) induction of microcracks in the grain,
		iv) detachment of the grain,
		v) detachment of grain fragments.
		
		b) Zhou and Bahadur (1993), Butler (1989), Kato (1990) and Bhushan and Sibley (1981) suggest two theories:
		   i) purely elastic fracture mechanism: *Hertzian* conical cracks;
		   ii) elasto-plastic fracture mechanism: lateral and radial cracks.
		
		c) Finnie (1960) – glass eroded by large spherical particles: erosion by intersection of *Hertzian* conical cracks.
		
		d) Ritter *et al.*(1984) and Evans (1982) – alumina eroded by SiC: cleavage and intergranular fractures in a *pit*.
		
		e) Wiederhorn and Hockey (1983) – alumina eroded by SiC: pit formation.
		
		f) Morrison *et al.*(1985) – mullite eroded by alumina: formation of a central crater with radial and lateral cracks.
		
		g) Zhou and Bahadur (1993) – aluminas with different contents of glass and zirconia, and high-purity alumina eroded with SiC – intergranular fracture type without plastic deformation and radial cracks.
		
		h) Marques (2006) – aluminas with and without vitreous phase eroded by electrofused alumina: main mechanism involves formation of erosion pits, by microcracking at alumina grain boundaries.
			a) Marques (2006) aluminas with and without vitreous phase eroded by electrofused alumina - formation of pits with rounded edges and ductile fracture.
			
			b) Zhou and Bahadur (1995) – alumina+glass eroded by SiC at 800ºC: erosion by plastic deformation, rounding of the eroded edge and brittle fracture.

Table 6 (*continued*)

Material	Angle	Temperature	
		Low (25ºC)	**High**
Bulk *cermet*	Low	a) Hussainova *et al.* (2001) –impact angle45º: *cermet* 85Cr$_3$C$_2$15Ni eroded by SiC: erosion by lateral cracking and decrusting. b) Hussainova *et al.* (2001) – impact angle 45º: *cermet* 85Cr$_3$C$_2$15Ni eroded by SiO$_2$: smooth and rounded morphology, detachment of carbide grains and consequent detachment of the metallic matrix. c) Bergman *et al.* (1997) – – impact angle 45º: high-speed steel with primary and secondary carbides eroded by SiO$_2$: erosion by cumulative indentation resulting in micro-fatigue and material loss.	
	High	a) Wayne *et al.* (1989) –adapted model from Evans *et al.*(1976) suggests the mechanisms: i) indentation by the particles; ii) sub-superficial development of lateral cracks, propagation to the surface, iii) micro-delamination and material removal. b) Beste *et al.* (2001) –TiC, TaNbC and WC carbides (*binderless*), with 0,25% Co and WC size of 5μm eroded by SiC: sites of local plastic deformation and removal by intergranular brittle fracture. c) Beste *et al.* (2001) –TiC, TaNbC and WC carbides (*binderless*), with 0,25% Co and WC size of 2μm eroded by SiC: plastic deformation and removal by intergranular brittle fracture outside the localarea of impact l. d) Beste *et al.* (2001) –TiC, TaNbC and WC carbides (*binderless*), with 0,25% Co and WC size of 0,6μm eroded by SiC: dominant erosion by plastic deformation, with small contribution of brittle fracture. e) Beste *et al.* (2001) –TiC, TaNbC and WC carbides (*binderless*), with 0,25% Co and WC size of 5μm eroded by SiC: erosion by intergranular brittle fracture with small traces of plastic deformation.	

Table 6 (*continued*)

Material	Angle	Temperature	
		Low (25ºC)	**High**
Coating cermets	Low	a) Levy *et al.* (1983) *apud* Levy and Wang (1988) – Cr_3C_2 by *D-gun* eroded by a mixture of oxides ($38\%Cr_2O_3$, $24\%Fe_2O_3$, $16\%MgO$, $15\%Al_2O_3$ and $7\%SiO_2$): erosion similar to polishing, smooth surface and some incrustations of particles. b) Levy *et al.* (1983) *apud* Levy and Wang (1988) – WC-Co by plasma *spray* eroded by a mixture of oxides ($38\%Cr_2O_3$, $24\%Fe_2O_3$, $16\%MgO$, $15\%Al_2O_3$ and $7\%SiO_2$): erosion process propagates over the eroded material and fills the voids. c) Kulu *et al.* (2005) – WC-17Co by *HVOF* eroded by rounded SiO_2 particles with porosity less than 3%: micro-cutting processes and in the case of coatings with low content of metallic phase, the process of low cycle fatigue and fracture occurs.	a) Wang and Shui (2003) – WC-17CoCr or Cr_3C_2-NiCr both *blended*, eroded by mineral coal ash at 300ºC: cracked and decrusted morphology, indicating mechanisms of brittle erosion. b) Wang and Verstak (1999) – Cr_3C_2/TiC-NiCrMo eroded by angular particles of mineral coal ashes, at temperature of 300ºC and velocity of 60 m/s: erosion by cracking and decrusting, presence of radial and lateral cracks. c) Kulu *et al.* (2005) – NiCrSiB eroded by SiO_2 rounded particles at temperature of 700ºC: erosion by ploughing.
	High	a) Levy *et al.* (1983) *apud* Levy and Wang (1988) – Cr_3C_2 by *D-gun* eroded by a mixture of oxides ($38\%Cr_2O_3$, $24\%Fe_2O_3$, $16\%MgO$, $15\%Al_2O_3$ and $7\%SiO_2$): erosion by cracking and decrusting of small pieces. b) Hawthorne *et al.*(1999) – Ni-based alloys eroded by Al_2O_3: cutting and *platelet* mechanisms. c) Hawthorne *et al.*(1999) – WC-12Co eroded by Al_2O_3: erosion by cutting, formation of *platelets,* and occasionally removal of individual carbides. d) Kulu *et al.* (2005) – NiCrSiB by *HVOF* eroded by rounded SiO_2 particles with porosity less than 3%: carbide fracture or removal of micro-particles. e) Kulu *et al.* (2005) – WC-17Co by *HVOF* by rounded SiO_2 particles with porosity less than 3%: carbide fracture or removal of micro-particles.	a) Wang and Verstak (1999) – Cr_3C_2/TiC-NiCrMo eroded by angular particles of mineral coal ashes, at temperature of 300ºC and velocity of 60 m/s: erosion by formation of small cracks and decrusting with incrustation of erodents. b) Kulu *et al.* (2005) – NiCrSiB eroded by rounded particles of SiO_2 at temperature of 700ºC: erosion by ploughing. c) Matthews, James and Hyland (2009) - HVAF coating 75 Cr_3C_2-25(Ni(20Cr)) eroded by crushed alumina (nominalsizeof 20–25 μm), at temperature of 700ºC (velocity of 225 m/s) and 800ºC (velocity of 225 m/s): erosion by localized mass loss (fracture of flakes and platelets) and extensive mass loss (gross chipping mechanism) d) Matthews, James and Hyland (2009) - HVOF coating 75 Cr_3C_2-25(Ni(20Cr)) eroded by crushed alumina (nominalsizeof 20–25 μm), at temperature of 700ºC (velocity of 225 m/s) and 800ºC (velocity of 225 m/s): erosion less ductile than in the HVAF coating with less ductile deformation in some single impact indents around the periphery of the steady state erosion crater.

Table 7 Summary of erosive wear mechanisms observed on $NiCr-Cr_3C_2$ coatings.

Material		T (ºC)	Angle	
			Low (30º)	**High (90º)**
NiCr		**25 a 800**	Ductile wear (by cutting).	Ductile wear (*platelets* formation).
Cr_3C_2		**25**	Brittle (angular areas, sharp corners and small cracks, as well *pits*).	
		200 a 800	Brittle (Sharp corners, small cracks and the presence of *pits*), slight rounding of the edges as the temperature increases.	
$NiCr-Cr_3C_2$	**High porosity (*Plasma spray*)**	**25**	$NiCr35CrC_{29\%}$: ductile, similar to NiCr.	$NiCr35CrC_{29\%}$: brittle, similar to Cr_3C_2.
			$NiCr70CrC_{31\%}$: brittle, similar to $CrC_{28\%}$	$NiCr70CrC_{31\%}$: brittle, similar to $CrC_{28\%}$.
		200 a 800	$NiCr35CCr_{29\%}$: Ductile wear (by cutting), but in less amount than NiCr.	$NiCr35CCr_{29\%}$: brittle, (sharp corners, craters and cracks).
			$NiCr70CCr_{31\%}$: brittle (sharp corners and cracks).	$NiCr70CCr_{31\%}$: brittle (sharp corners, cracks and *pits*).
	Low porosity (*HVOF*)	**25**	Ductile wear (by cutting), losing dimensions as increase the carbide content.	$NiCr30CrC_{3\%}$: ductile as *platelets*, but in less intensity than NiCr coating NiCr. Signs of brittle wear (angular areas and cracks increasing with Cr_3C_2).
		200 a 800	$NiCr35CCr_{3\%}$: Ductile wear (by cutting).	$NiCr35CCr_{3\%}$: ductile as *platalets*, decrease in the intensity with the temperature.
			$NiCr70CCr_{3\%}$: Ductile wear (by cutting).	$NiCr70CCr_{3\%}$: Modification of morphology of ductile wear as platelet, from 200ºC, where ductile wear by cutting appears.
			$NiCr75CCr_{3\%}$: Ductile wear (by cutting).	$NiCr75CCr_{3\%}$: ductile as *platelets*.

7 Final Considerations

Considering the extensive research presented in this review, an agreement was noted among authors concerning the erosive wear of metallic (ductile) and ceramic (brittle) materials. In the first group, authors point out to a wear caused by plastic deformation, by decrusting, cutting, or platelet formation from the surface. Otherwise in brittle materials, such as ceramics, the removal occurs by interlinking of microcracks (lateral and radial) which diverge from the strike point of the particle against the surface, as well as by pit formation.

Regarding the erosion of bulk cermet materials, the authors suggest that the degradation is very complex, due to the nonhomogeneous character of the material, involving behaviors ranging from dominant plasticity to dominant brittle fracture. The type of damage caused on the material, according to the investigated authors, depends on the system tribology: i) target material: carbides size, type of carbide material, type of bonding phase material, and physical and mechanical properties of the cermet. ii) erodent material: morphology, size, and mechanical properties, and iii) erosion tests parameters: velocity, rate of erosion, attack angle, and type of test. Generally, the authors state that the erosion of these materials starts in the bonding phase (ductile matrix): the carbide grains lose their linkage to the matrix, essential for maintaining the integrity of the cermet, and the eroded surface leads to exposure of the carbides, which are soon detached. However, the mode as the removal of the bonding phase or the carbides occurs is still an open subject, some theories were proposed, but there is no agreement among them.

Pawlowski (2008) *apud* Lansdown and Price (1986) suggest a simplified guide for the selection of a coating material, depending on erodent speed and impact angle (Figure 56). However, this is a very simplified diagram, because the erosion wear depends on many factors and mainly on the microstructure of the target material and erodent particles. Kulu and Pihl (2002) suggest some rules for coatings to be resistant against erosion:

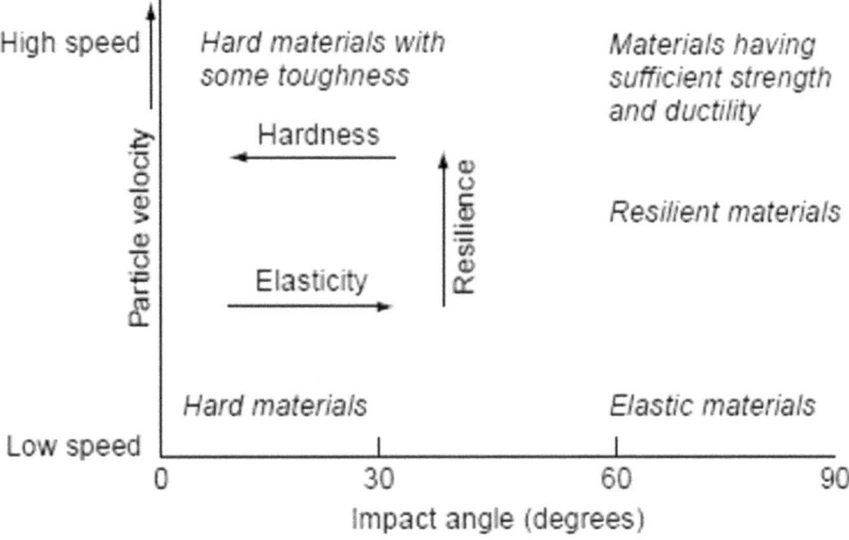

Fig. 56 Simplified guide for selecting materials which resist erosion by particles having different speeds and impact angles (PAWLOWSKI (2008) *apud* LANSDOWN and PRICE (1986).

i) minimum porosity – high-velocity spray processes should be used and the porosity should be kept lower than 3 %.

ii) optimal hardness of coatings – the hardness depends on the erosion conditions. At small impact angles, the hardness must be high and indeed higher than that of the erodent. At higher angles, the hardness should be optimized.

iii) optimal microstructure – at oblique impact, the cermet structure seems to be optimal. At normal impact, dispersion strengthened the metal or composite with such a matrix.

iv) hardness of abrading – the erosion resistance is guaranteed if the hardness of the erodent is lower than the hardness of the coating.

In fact, in the erosive wear of cermet material coatings, it can be observed that the authors do not reach a consensus among their conclusions, they only suggest some features about the target material. Some aspects are coincident and some are absolutely contradictory. Indeed, this is due to the fact that the erosive wear of these materials is related to many factors which depend on the involved system tribology and mainly on the different microstructures of the target material, lamellar. On the one hand, a common point among the researchers can be mentioned: cermet coatings used under wear condition in which a mixture of mechanisms are involved, attack angles and wear modes, due to the combination of the physical and mechanical properties of ceramic and metals, such as the high melting point and high chemical and thermal stabilities of carbides. On the other hand, the authors point out the need of testing different coatings under distinct erosive wear conditions, because there is a lack of information about the behavior of different classes and types of cermets. There is almost consensus regarding the microstructure of coatings that are suitable to high-temperature erosion: lamellas and small grain size, low porosity, low amount of oxides, good adherence, and absence of cracks, which implies the type of cermet used and the type of application technique. However, regarding high temperatures, few authors mention the importance of porosity, for the oxide formation, which can protect the coating, and the temperatures at which the coating can be used satisfactorily. Furthermore, the possible modification of the mechanical properties of the eroded surface, due to the impact of erodent particles, usually is not mentioned, but can occur, modifying the mechanism responsible for the wear of these coatings. Vicenzi (2007) summarizes the dependencies of the erosion strength of cermet coatings: factors associated with the microstructure of the coating (nature - metallic/ceramic/pores, proportion, shape, orientation, size and distribution of present phases) and operational variables of the erosion process, such as, for example, erodent attack angle and test temperature. The phenomena acting, in combination or not, depend on temperature: i) modification of mechanical properties (hardness and modulus of elasticity); ii) release of residual stress with or without crack propagation inside the coating and on the interface coating/substrate; iii) incrustation of erodent particles and reactions with oxygen (formation of oxides/liberation of CO/CO_2).

References

ASTM G40-92. Annual Book of ASTM Standards. v. 3.02, p.160 (1992)

Ball, A.: The mechanisms of wear, and the performance of engineering materials. J.S. Afr. Inst. Min. Metall 86(1), 1–13 (1986)

Barbezat, G., Nicol, A.R., Sickinger, A.: Abrasion, erosion and scuffing resistance of carbide and oxide ceramic thermal sprayed coatings for different applications. Wear 162 - 164, 529–537 (1993)

Berger, L.M., et al.: Thermal Spray: Practical Solutions for Engineering Problems (1996)

Bergman, F., Hedenqvist, P., Hogmark, S.: The influence of primary carbides and test parameters on abrasive and erosive wear of selected PM high speed steels. Tribology International 30, 183–191 (1997)

Berthier, Y.: Experimental evidence for friction and wear modeling. Wear 139, 77–92 (1990)

Beste, U., et al.: Particle erosion of cemented carbides with low Co content. Wear 250, 809–881 (2001)

Bhushan, B., Sibley, L.B.: Silicon nitride rolling bearing for extreme operation conditions. ASLE Transactions 25, 417–428 (1981)

Butler, E.G.: Engineering ceramics: Aplications and testing requirements. In: Dyson, B.F., Lohr, R.D., Morrel, R. (eds.) Mechanical testing of engineering ceramics at high temperatures, pp. 1–10. Elsevier Applied Science, New York (1989)

Cousens, A.K., Hutchings, I.M.: In: Field, J.F., Corney, N.S. (eds.) Proc. 6th Int. Conf. on Erosion by Liquid and solid impacts, Cavendish Laboratory, Cambridge University (1983)

Ctibor, P., Lechnerova, R., Benes, V.: Quantitative analysis of pores of two types in a plasma-sprayed coating. Materials Characterization 56, 297–304 (2006)

D'Errico, E.G., et al.: A study of cermets' wear behavior. Wear 203 - 204, 242–246 (1997)

Engqvist, H., Axén, N., Hogmark, S.: Tribological properties of a binderless carbide. Wear 232, 157–162 (1999)

Evans, A.G.: Structural Reability: A Processing dependent phenomenon. Journal of American Ceramic Society 65(3), 127–137 (1982)

Evans, A.G., Wilshaw, R.: Quasi-static solid particle damage in brittle solids - I. Observations, analysis and implications. Acta Metali. 24, 939 (1976)

Feng, Z., Ball, A.: The erosion of four materials using seven erodents - towards an understanding. Wear 233-235, 674–684 (1999)

Finnie, I.: Erosion of surfaces by solid particles. Wear 3, 87–103 (1960)

Finnie, I.: Some reflections on the past and future of erosion. Wear 186-187, 1–10 (1995)

Gou, D.Z., Wang, L.J., Li, J.Z.: Erosive wear of low chromium white cast iron. Wear 161, 173–178 (1992)

Gross, K.: Thermal Spraying – An Introduction in,
http://www.azom.com/Details.asp?ArticleID=542
(accessed in April 3, 2011)

Hawthorne, H.M., et al.: Comparison of slurry and dry erosion behaviour of some HVOF thermal sprayed coatings. Wear 225 - 229, 825–834 (1999)

Hidalgo, V.H., et al.: A comparative study of high-temperature erosion wear of plasmasprayed NiCrBSiFe and WC–NiCrBSiFe coatings under simulated coal-fired boiler conditions. Tribology International 34, 161–169 (2001)

Hoppert, S.: Alumina ceramics – Superior materials for protection against wear and corrosion. Siegburg, 3–11 (1989)

Hussainova, I., Kubarsepp, J., Pirso, J.: Mechanical properties and features of erosion of cermets. Wear 250, 818–825 (2001)

Hutchings, I.M.: A model for the erosion of metals by spherical particles at normal incidence. Wear 81, 269–281 (1981)

Hutchings, I.M.: Deformation of metal surfaces by the oblique impact of square plates. International Journal of Mechanical Science 19, 45–52 (1977)

Hutchings, I.M.: Mechanical and metallurgical aspects of the erosion of metals. Wear, 393–427 (1979)

Hutchings, I.M.: Normal impact of metal projectiles against a rigid target at low velocities. International Journal of Mechanical Science 23, 255–261 (1981)

Hutchings, I.M.: Thermal effects in the erosion of ductile metals. Wear 131, 105–121 (1989)

Hutchings, I.M., Winter, R.E., Field, J.E.: Proc. Roy. Soc. London A348, pp. 379–392 (1976)

Kato, K.: Tribology of ceramics. Wear 136, 117–133 (1990)

Kulu, P.: The Abrasive Erosion Resistance of Powder Coatings. J. Tribologia: Finnish J. Tribology 4, 12–25 (1989)

Kulu, P., Hussainova, I., Veinthal, R.: Solid particle erosion of thermal sprayed coatings. Wear 258, 488–496 (2005)

Kulu, P., Pihl, T.: Selection Criteria for Wear Resistant Powder Coatings Under Extreme Erosive Wear Conditions. Journal of Thermal Spray Technology - ASM International 11, 517–522 (2002)

Kulu, P., Pihl, T., Halling, J.: Wear-resistant WC–Co–NiCrSiB composite coatings. In: Proceedings of the Eighth International Conference of Tribology, Aarhus, pp. 809–817 (1998)

Kulu, P., Tümanok, A., Zimakov, S.: Treatment of Recycled Hardmetals. In: Book of Proceedings of 4th ASM International Conference and Exhibition on the Recycling of Metals, pp. 319–327. ASM International, Vienna (1999)

Kulu, P., Veinthal, R.: Wear Resistance of High Velocity Thermal Sprayed Coating. In: Proceedings of 9th Nordic Symposium on Tribology, NORDTRIB 2000, Technical Research Centre of Finland (VTT), Espoo, Finland, pp. 87–95 (2000)

Kulu, P., Veinthal, R., Kõo, J., Lille, H.: Mechanism of Abrasion ErosionWear of Thermal Sprayed Coatings.Advances in Mechanical Behaviour, Plasticity and Damage. In: Proceedings of EUROMAT 2001, pp. 651–656. Elsevier, New York (2001)

Kulu, P., Zimakov, S.: Wear resistance of thermal sprayed coatings on the base of recycled hardmetal. Surface and Coatings Technology 130, 46–51 (2000)

Kulu, P., Zimakov, S.: Wear-resistant composite coatings. In: Surface Engineering, EUROMAT 1999, vol. 11, pp. 144–149. Wiley-VCH, Weinheim (2000)

Kunioshi, C.T., Correa, O.V., Ramanathan, L.V.: Comportamento de oxidação e erosão-oxidação de revestimentos HVOF a base de NiCr. In: XVI CBECIMAT, Porto Alegre, Brasil, pp. 1–16 (2004)

Lansdown, A.R., Price, A.L.: Materials to Resist Wear – A Guide to Their Selection and Use. Pergamon Press, Oxford (1986)

Larsen-Basse, J.: Effect of composition, microstructure and service conditions on the wear of cemented carbides. J. Met. 11, 35–41 (1983)

Laugier, S., Hyland, M., James, B.: Acta Mater. 51, 4267 (2003)

Levy, A.V., Chik, P.: The effects of erodent composition and shape on the erosion of steel. Wear 89, 151–162 (1983)

Levy, A.V., Wang, B.Q.: Erosion of hard material coating systems. Wear 121, 325–346 (1988)

Levy, A.V., Yau, P.: Erosion of steels in liquid slurries. Wear 98, 163–182 (1984)

Lewis, T., Sokol, L., Hanna, E.: Optimization of Gator-Gard applied chrome carbide-McrAlY composite overlays for maximum solid particle erosion resistence. In: ASM National Thermal Spray Conference Proceedings, Orlando, FL, p. 149 (1989)

Lille, H., Kõo, J., Kulu, P., Pihl, T.: Residual stresses in different thermal spray coatings. Proceedings of the Estonian Acad. Sci. Eng. 8, 162–173 (2002)

Madruga, T.P., Bergmann, C.P., Silveira, M.M.: Resistência ao desgaste de aluminas. In: Anais do 38° Congresso Brasileiro de Cerâmica, vol. 1, pp. 198–203 (1994)

Mann, B.S., Arya, V.: Abrasive and erosive wear characteristics of plasma nitriding and HVOF coatings: their application in hydro turbines. Wear 249(5-6), 354–360 (2001)

Marques, C.M.: Relação entre microestrutura e desgaste erosivo a frio e a quente em materiais cerâmicos à base de alumina. In: Tese (Doutorado), Programa de Pós-graduação em Engenharia de Minas, Metalúrgica e de Materiais (PPGEM), Universidade Federal do Rio Grande do Sul, Porto Alegre (2006)

Matthews, S., Hyland, M., James, B.: J. Therm. Spray Technol. 13, 526 (2004)

Matthews, S., James, B., Hyland, M.: High temperature erosion of Cr3C2-NiCr thermal spray coatings — The role of phase microstructure. Surface & Coatings Technology 203, 1144–1153 (2009)

Morrison, C.T., Routbort, J.L., Scattergood, R.O.: Solid particle erosion of mullite. Wear 105, 19–27 (1985)

Powlowski, L.: The Science and Engineering of Thermal Spray Coatings. Wiley, New York (1994)

Praxair. Disponível em . Acesso em fevereiro de (2006), http://www.praxair.com

Raak, E.: Erosion Wear in Coal Utilization, Hemisphere, Washington, DC (1988)

Reshetnyak, H., Kübarsepp, J.: Mechanical properties of hard metals and their erosive wear resistance. Wear 177, 185–193 (1994)

Ritter, J.E., Davidge, R.W.: Strength and its variability in ceramics with particular reference to alumina. J. Am. Ceramic Soc. 67(6), 432–437 (1984)

Roy, M., Ray, K.K., Sundararajan, G.: An analysis of the transition from metal erosion to oxide erosion. Wear 217, 312–320 (1998)

Scrivani, A., et al.: A contribution to the surface analysis and characterization of HVOF coatings for a petrochemical application. Wear, 107–113 (2001)

Shanov, V., Tabakoff, W.: Erosion resistance of coatings for metal protection at elevated temperatures. Surface and Coating Technology, 88–93 (1996)

Sheldon, G.L., Finnie, I.: Trans. ASME. 88B, 387 (1966)

Sheldon, G.L.: Similarities and differences in the erosion behavior of materials. J. Basic Eng. 92, 619–626 (1970)

Shetty, D.K., Wright, I.G., Clauer, A.H.: Effects of composition and microstructure on the slurry erosion of WC-Co cermets. Wear 114, 1–18 (1987)

Shewmon, P., Sundararajan, G.: The erosion of metals. Ann. Rev. Mater. Sci. 13, 301 (1983)

Shipway, P.H., Hutchings, I.M.: The role of particle properties in the erosion of brittle materials. Wear 193, 105–113 (1996)

Söderberg, S., et al.: Erosion classification of materials using a centrifugal erosion tester. Tribology International 14, 333–343 (1981)

Souza, V.A.D., Neville, A.: Linking electrochemical corrosion behaviour and corrosion mechanisms of thermal spray cermet coatings (WC_/CrNi and WC/ CrC_/CoCr). Materials Science and Engineering A 352, 202–211 (2003)

Stack, M.M., Peña, D.: Solid particle erosion of Ni-Cr/WC metal matrix composites at elevated temperatures: construction of erosion mechanism and process control maps. Wear, 489–497 (1997)

Stein, K.J., Schorr, B.S., Marder, A.R.: Erosion of thermal spray MCr–Cr$_2$C$_3$ cermet coatings. Wear 224, 153–159 (1999)

Stoicaa, V., et al.: Sliding wear evaluation of hot isostatically pressed (HIPed) thermal spray cermet coatings. Wear 257, 1103–1124 (2004)

Sue, J.A., Tucker Jr., R.C.: Surface & Coatings Technology, vol. 32, p. 237 (1987)

Tabakoff, W.: Experimental study on the effects of specimen sizes on erosion. Wear 86, 65–72 (1983)

Toma, D., Brandl, W., Marginean, G.: Wear and corrosion behaviour of thermally sprayed cermet coatings. Surface and Coatings Technology 138, 149–158 (2001)

Tu, J.P., Mao, Z.Y., Li, J., Wang, L.Z.: Erosion Behavior of a Thermally Sprayed Ni–WC Coating at High Temperature. In: Proceedings of the Fourth National Thermal Spray Conference, Pittsburgh, PA, p. 53 (1991)

Upadhyaya, G.S.: Materials science of cemented carbides - an overview. Materials and Design 22, 483–489 (2001)

Vicenzi, J.: Relação entre microestrutura e erosão (a frio e a quente) de revestimentos do sistema NiCr-Cr$_3$C$_2$ obtidos por aspersão térmica. In: Tese (Doutorado), Programa de Pós-graduação em Engenharia de Minas, Metalúrgica e de Materiais (PPGEM), Universidade Federal do Rio Grande do Sul, Porto Alegre (2007)

Voyer, J., Marple, B.R.: Sliding wear behavior of high velocity oxy-fuel and high power plasma spray-processed tungsten carbide-based cermet coatings. Wear 229, 135–145 (1999)

Walsh, D.W.: Effects of cyclic heating on diffusion bonding of steel powders. Metal Powder Report 7, 58 (1992)

Walsh, P.N., Tabakoff, W.: Comparative erosion resistance of coatings intended for steam turbine components. Power 10, 1 (1990)

Wang, B.Q., Lee, S.W.: Elevated temperature erosion of several thermal-sprayed coatings under the simulated erosion conditions of in-bed tubes in a fluidized bed combustor. Wear 203-204, 580–587 (1997)

Wang, B.Q., Seitz, M.W.: Comparison in erosion behavior of iron –based coatings sprayed by three different arc-spray processes. Wear, 755–761 (2001)

Wang, B.Q., Shui, Z.R.: Hot erosion behavior of carbide–metal composite coatings. Journal of Materials Processing Technology (2003) (in press)

Wang, B.Q., Verstak, A.: Elevated temperature erosion of HVOF Cr$_3$C$_2$/TiC-NiCrMo cermet coating. Wear 233-235, 342–351 (1999)

Wang, B.Q., Luer, K.: In: Berndt, C.C., Sampath, S. (eds.) Thermal Spray Industrial Applications, p. 115. ASM International, Materials Park (1994)

Wang, B.Q., Geng, G.Q., Levy, A.V.: Erosion–corrosion of thermal spray coatings. Surface and Coatings Technology 43, 859 (1990)

Wayne, S.F.E., Buljan, T.: Wear and design of ceramic cutting tool materials. Wear 133, 309–321 (1989)

Wensink, H., Elwenspoek, M.C.: A closer look at the ductile–brittle transition in solid particle erosion. Wear 253, 1035–1043 (2002)

Wiederhorn, S.M., Hockey, B.J.: Effect of materials parameters on the erosion resistance of brittle materials. J. Mat. Sci. 18, 766–780 (1983)

Yao, J., Zhang, B.Z., Fan, J.R.: An experimental investigation of a new method for protecting bends from erosion in gas-particle flows. Wear 240, 215–222 (2000)

Zhang, X.C., Xu, B.S., Tu, S.T., Xuan, F.Z., Wang, H.D., Wu, Y.X.: Fatigue resistance and failure mechanisms of plasma-sprayed CrC–NiCr cermet coatings in rolling contact. International Journal of Fatigue 31, 906–915 (2009)

Zhou, J., Bahadur, B.: Erosion characteristics of alumina ceramics at high temperatures. Wear 181-183, 178–188 (1995)

Zhou, J., Bahadur, B.: Sem studies of material damage in alumina ceramics by angular single and multiple particle impacts. Wear 164, 285–295 (1993)

Zhu, S., Xu, B.S., Yao, J.K.: High quality ceramic coatings sprayed by high efficiency hypersonic plasma spraying gun. Mater. Sci. Forum 3981- 4, 475–479 (2005)

MIX
Papier aus verantwortungsvollen Quellen
Paper from responsible sources
FSC® C105338

If you have any concerns about our products,
you can contact us on
ProductSafety@springernature.com

In case Publisher is established outside the EU,
the EU authorized representative is:
Springer Nature Customer Service Center GmbH
Europaplatz 3, 69115 Heidelberg, Germany

Printed by Libri Plureos GmbH
in Hamburg, Germany